SpringerBriefs in Philosophy

More information about this series at http://www.springer.com/series/10082

Hassan Tahiri

Mathematics and the Mind

An Introduction into Ibn Sīnā's Theory of Knowledge

Springer

Hassan Tahiri
Meknès
Morocco

ISSN 2211-4548 ISSN 2211-4556 (electronic)
SpringerBriefs in Philosophy
ISBN 978-3-319-25236-0 ISBN 978-3-319-25238-4 (eBook)
DOI 10.1007/978-3-319-25238-4

Library of Congress Control Number: 2015951774

Springer Cham Heidelberg New York Dordrecht London

Printed on acid-free paper

Springer International Publishing AG Switzerland is part of Springer Science+Business Media (www.springer.com)

Preface

My first contact with the Arabic tradition and the work of Ibn Sīnā, better known in the West as Avicenna and often referred to as the 'physician', goes back to the publication of the volume *The Unity of Science in the Arabic Tradition* in 2008 of which I was a coeditor. Ibn Sīnā got the lion's share; the majority of contributors chose to write about the author of *al-Shifā'*. I learned a great deal from the edition of this volume, and in particular, I became aware of the importance and complexity of history. My subsequent research focused on Ibn Sīnā's scientific works.

One of the outcomes of the publication of the volume was the deepened understanding of Ibn Sīnā's logical and mathematical thought. This was mainly the work of leading logicians and mathematicians such as Roshdi Rashed, Wilfrid Hodges, Shahid Rahman and his collaborators who have demonstrated in recent papers how far Ibn Sīnā advanced logical research by attempting to develop a synthetic logical approach that combined logic of quantification, logic of identity or equivalence and a theory of inference.

The present work is complementary to the latest research on Ibn Sīnā's scientific works, and it concentrates on another feature of his thought: the dynamic interaction between his epistemic attitude and scientific practice.

More precisely his reflections on the basic concepts of mathematics and logic have led him to develop a basic epistemic standpoint, the results of which are then used to develop them further, and reflection on the further development of both disciplines in turn increases the understanding of the underlying epistemic acts.

This work has been developed in the context of the research project "Argumentation and Scientific Change: A case Study of How Ibn al-Haytham's *al-Shukūk* Changed the Course of Astronomy Forever" sponsored by the Portuguese foundation of science and technology (Fundação para a Ciência e Tecnologia). The research was conducted in the Centre for Philosophy of Science of the University of Lisbon (CFCUL). I would like to thank the members of the Centre for their help and support, its head Prof. Olga Pombo and Dr. Zbigniew Kotowicz who very kindly read the final manuscript and with whom I had a number of elucidating and fruitful discussions on the subject of this study. My special thanks go to an anonymous reviewer for his helpful comments and Ms. Lue Christi of Springer for her great assistance throughout the edition of this work. Most of all, I am particularly grateful to Shahid Rahman and Roshdi Rashed, both eminent experts respectively in philosophy and history of logic and philosophy and history of mathematics, for the time they took to read previous drafts and for their invaluable suggestions which significantly improved the manuscript.

CFCUL, Lisbon Hassan Tahiri
June 2015

Contents

About the Book

This book examines how epistemology was reinvented by Ibn Sīnā, one of the influential philosopher-scientists of the classical Islamic world who was known to the West by the Latinised name Avicenna. It explains his theory of knowledge in which intentionality acts as an interaction between the mind and the world. This, in turn, led Ibn Sīnā to distinguish an operation of intentionality specific to the generation of numbers.

The author argues that Ibn Sīnā's transformation of philosophy is one of the major stages in the de-hellinisation movement of the Greek heritage that was set off by the advent of the Arabic-Islamic civilisation. Readers first learn about Ibn Sīnā's unprecedented investigation into the concept of the number and his criticism of such Greek thought as Plato's realism, Pythagoreans' empiricism and Aristotle's conception of existence.

Next, coverage sets out the basics of Ibn Sīnā's theory of knowledge needed for the construction of numbers. It describes how intentionality turns out to be key in showing the ontological dependence of numbers as well as even more critical to their construction.

In describing the various mental operations that make mathematical objects intentional entities, Ibn Sīnā developed powerful arguments and subtle analyses to show us the extent our mental life depends on intentionality. This monograph thoroughly explores the epistemic dimension of this concept, which, the author believes, can also explain the actual genesis and evolution of mathematics by the human mind.

Chapter 1
Introduction

The Reinvention of Knowledge

Few philosophers that have been studied as much as Ibn Sīnā have been as much misunderstood. His extraordinary ability to reflect upon and write in a variety of styles about seemingly every topic in every domain has steered his thought from philosophy and theology to mysticism and esoterism. Instead of helping us to learn and understand better Ibn Sīnā than he has previously been understood, the recent surge of Avicennan studies only adds more confusion to the already complex social context which he was living in. Further evidence of his creativity is that he knew how to make his own life more legendary by writing an autobiography but which only deals with the first part of his life and entrusting to one of his disciples the task of finishing the rest of his biography. His life turns out to be so fascinating that it gave rise to two popular novels which appeared around the same time.[1] It is not surprising that many scholars have been struggling for years to separate fact from fiction. Ibn Sīnā's philosophical thought is as open to speculation as is his autobiography, with many scholars regarding him as Aristotelian, others as Neo-Platonist, a third group consider his philosophy as a mixture of Aristotelianism and Neo-Platonism, and a fourth group portray it as an extension of late

[1] *The Physician* by Noah Gordon (New York, Fawcett Crest, 1986) and *Avicenne ou La route d'Ispahan* by Gilbert Sinoué (Paris, Denoël, 1989).

H. Tahiri, *Mathematics and the Mind*,
SpringerBriefs in Philosophy, DOI 10.1007/978-3-319-25238-4_1

Alexandrian Aristotelianism.[2] By bringing him back to an ancient tradition, however, they fail to recognise the epistemic obstacles inherited by his own tradition, the new questions generated by his own scientific context and to appreciate the new means he invented to deal with them.

The case of Ibn Sīnā is by no means an exception; it is in fact part of a general misconception of the history of Arabic-Islamic philosophy. This deep entrenched tendency that systematically links Arabic-Islamic philosophers backwards to the Greek tradition, despite its interest in describing the background against which their work arose, overshadows the underlying process at work and the real scope of their undeniable achievements and innovations. The various Greek affiliations to which they are linked amounts in fact to the Hellenisation of their thoughts as if the Arabic-Islamic civilisation is part of the Greek world view, while what was happening was something far more important for the history and development of thought, namely, the universalisation of science and philosophy or more generally the trans-cultural trans-national character of knowledge. This aspect of the worldwide historical development of science and philosophy is outshined by their current periodisation[3] which gives the misleading impression that the so-called "globalisation" of knowledge is a new western phenomenon (see Rashed (1978) for more on how science is unanimously peculiarly perceived and portrayed as a western phenomenon). This is particularly relevant for the study of the Arabic tradition for it is during the Arabic period when the first globalisation of knowledge was achieved. The wide circulation of scientific and philosophical books from Samarkand in the

---

[2]In his pioneering paper "Philosophy of mathematics" (in Rahman et al. 2008), Roshdi Rashed makes crystal-clear in many passages that his mathematical investigations do not support the received view:

> To understand the distance put by Ibn Sīnā between himself and traditional, Hellenistic and Greek classifications as well as between himself and his own theoretical classification, it is worth introducing here one of his predecessors, al-Fārābī. (Rashed 2008, 167)
>
> In the presence of this new discipline which has to be taken into account, the new classification of the sciences which aimed at both universality and exhaustiveness has to justify in one way or another the abandonment of certain Aristotelian theses. Names such as "science of ingenious devices", "derivative parts" are coined so that a non-Aristotelian zone can be arranged within a received Aristotelian style of classification. The philosophical impact caused by such a revision is on a larger scale and—especially—more profound than mere taxonomic modification. (ibid., p. 168)
>
> It is sufficient to recall that, being neither Platonic nor Aristotelian, this new [formal] ontology arose, in part at least, due to the new results in mathematical sciences. (ibid., p. 169)

This distinguished historian of mathematics has set himself the task of demonstrating that what the received view finds it not so obvious: the various Greek affiliations have little intrinsic meaning in a scientific context based on a new concept of knowledge. For a general assessment about the present state of Arabic philosophical research, see his introduction to his paper.

[3]Despite its refutation since the second half of the last century following the discovery of the works of the Marāgha astronomers, it is unfortunate that the dogmatic periodisation imposed by modern historians of science has yet to be updated and revised by the scientific community. For more on this issue, see Tahiri (2008).

East to Toledo in the West has generated at least three more new traditions that did not exist before: the Persian, the Jewish and the rising powerful Christian tradition.[4] Many historians and scholars underestimate the unprecedented globalisation of knowledge that enables its non-stop development since the ninth century by confining the historical role of the Arabic tradition to a mere preserver and mediator of the Greek heritage. By taking for granted the evolution of science and philosophy from one culture to another, important epistemic questions about the significance and the timing of their emergence in a new culture are overlooked: why did it occur in the ninth century but did not happen before? Had the Greeks ever imagined for example that the philosophy they cultivated since the 6th century B.C. could be practiced one day in a language other than their own? More generally, how could science and philosophy be further developed if they remained encoded in a language which was no longer in use?

Epicurus (341–270 BC) is reported to have said that "only Greeks philosophise",[5] by this he probably means that philosophy is so specific to the Greek

---

[4]In the fourth chapter entitled "Le tribut des arabes avant le XIIIe siècle" of his third volume *Le Système du monde*, the great French historian Duhem points out that there is evidence which suggests that early translations were carried out before the millennium:

> Les premières infiltrations de la Science arabe en la Science latine sont, sans doute, très anciennes; il est fort probable qu'avant l'An Mil, les écoles de France possédaient déjà des livres traduits de ceux qu'avaient composés les sarrasins, et que certains de ces livres concernaient l'Astronomie pratique, celle qu'occupent la confection des tables et l'usage des instruments. (Duhem 1913, Tome III, p. 164)

He then reminds us of the turning point in western history brought about by the translation of Arab works

> Les livres traduits de l'Arabe vont se répandre, de plus en plus nombreux, dans les écoles latines; la science enseignée dans ces écoles n'en sera pas seulement accrue; elle sera, en même temps, orientée suivant une direction *toute différente* de celle qu'elle avait suivie jusqu'alors. (ibid., pp. 163–164; my emphasis)

[5]This quotation can be found in van den Bergh's second volume of his translation Abūl-Walīd ibn Rushd (1126–1198) or Averroes' *Tahāfut al-Tahāfut* (*The Incoherence of the Incoherence*), he cites two authorities as reference: Clement of Alexandria, *The Stromata*, I. 15 and Diogenes. Laertes X. 117. He immediately quotes Ibn Maimūn or Maimonides (1135–1204) to express what he has in mind: "one must know that everything the Moslems, *Mu'tazilites* as well as *Ash'arites*, have professed concerning these subjects, has been borrowed from the Greeks and Syrians who applied themselves to the criticism of the philosophers" (Maimonides, Dalālatu al-ḥā'irīn or *Guide for the Perplexed*, I. 71). The two quotations are put as epigraph to the first page of his substantial second volume (p. 219!) devoted entirely to notes and extensive commentaries in addition to a long introduction to his translation of the work of the philosopher of Cordoba. The two volumes are presented as proof of his claim and the definitive word on the topic. Misguided by such kind of prejudices still widely prevailing, our modern scholar fails to grasp the meta-dialogical issue that motivates Ibn Rushd's reaction which is the transformation of philosophy by our author. This is just another reminder that the philological approach used by scholars who work on the complex domain that lies between the Greek and Arab lands is by no means sufficient to fill the gaps in our understanding of the history of philosophy, and could even be misleading.

culture and the philosophical corpus accumulated over centuries is so enormous that its transmission to a different language is simply an impossible mission. He would be astonished by the worldwide practice of philosophy today and yet few scholars and historians know and recognise that this is the outcome of its transformation that goes back to the ninth century. How could have such a transformation been brought about? And why did it take such long time to materialise? If history is any indication, there was little awareness of the need for change during the immense life span of the Greek period. But what kind of change? And by what means could have it been brought about? Science and philosophy have been developed in the context of Greek culture to such extent that any further progress required their de-hellinisation; and by de-hellenisation we mean the substitution of Greek metaphysical and cultural conceptions that hampered the further progress towards a concept of knowledge that could cross frontiers and interact with different conceptual frames and traditions.

The significance of the Arabic-Islamic tradition lies in reinventing knowledge that set off the unstoppable de-hellinisation of the Greek heritage. The starting point for this new tradition that inaugurates the de-hellinisation era is its specific approach to knowledge which marks a major shift in the direction in which the scientific inquiry was practiced. For according to the Arabic-Islamic understanding, knowledge is and must be useful (العلم النافع); its usefulness lies in its ability to be at the service of society by yielding some practical benefits which in turn contribute to its further theoretical development.[6] By identifying itself with this conception of knowledge, Arabic-Islamic civilisation has conceived a distinctive and global project,[7] for its intellectuals, whether they are jurists or *mutakallimūn* (theologian-philosophers), poets or litterateurs, grammarians or linguists, scientists or artists, philosophers or mystics, who have all co-operated in its construction and

[6]In the introduction of his foundational work that appeared about 820, al-Khwārizmī (last decades of the 8th century–mid 9th century) explains that usefulness is what his Algebra is all about and mentions, after acknowledging his debt to the caliph for having encouraged him to compose his calculation of algebra and *al-muqābala*, some of the domains where his new science can significantly contribute to their development.

> I wanted it to include what is subtle in the calculation and what is the most noble in it, what people necessarily need in their inheritances, legacies, partitions, law-suits, and trade, and in all their dealings with one another where surveying, the digging of canals, geometrical computation are concerned, and other affairs relevant to calculation and to its sorts. (Al-Khwārizmī 2007, p. 94)

Like mathematics, astronomy, both theoretical and practical, was also developed in the same spirit, for more David King's *Astronomy in the Service of Islam* (1993). But the practical benefit goes beyond the material aspect of theoretical research. The usefulness of a scientific theory should obviously be understood in a wider sense that includes its capacity to interact with other scientific theories and contribute to their development, like for example the possible application of its concepts and forms of reasoning to another theoretical, empirical or social discipline.

[7]See Rosenthal's *Knowledge Triumphant* (1970), for how Islamic civilisation makes knowledge the basis of its foundation and development.

globalisation. The de-hellinisation of science and philosophy was systematic, wide ranging, took different forms and affected the Greek corpus in various ways. But it was not straightforward not without difficulties and not free from controversies. Works such as al-Kindī's *Philosophy Can Only Be Acquired Through the Mathematical Discipline*, al-Rāzī's *Doubts About Galen*, al-Fārābī's *Enumeration of the Sciences*, Ibn Sīnā's *Logic of the Orientals*, Ibn al-Haytham's *Doubts About Ptolemy*, al-Ghazālī's *The Incoherence of the Philosophers*, Ibn Bājja's doubts about the Aristotelian metaphysical conception of motion,[8] Ibn Taymiyya's *Against the Greek Logicians* are just some of the most striking examples that illustrate the enormous efforts spent to overcome the epistemic barriers erected by the Greek conception of science and philosophy.

Three new fundamental disciplines which were developed in the context of Arabic-Islamic culture have created the new social and scientific milieu needed for initiating the irreversible de-hellinisation movement. Islamic jurisprudence is the first scientific discipline to be set up by the jurists themselves. The formal codification of principles and rules, that structured the Arabic-Islamic society, was not achieved however without disagreement. The jurists agreed on the principles and the rules of inference to generate more laws but acknowledged the differences of their approaches to the interpretation of derivative laws and their applications. They then worked out a dynamic instrument that helped them move forward by developing rules and procedures to increase the rationality and efficiency of empirical debates; this gives rise to a new discipline called the *ādāb al-jadal* or the Rules of

---

[8]This is how Ibn Rushd describes Ibn Bājja's (1095–1138) double attack on Aristotle's conception of motion in the famous passage historically known as Comment 71:

> Ibn Bājja raises doubts in this passage in two places. One occurs when he says that the ratio of motion to motion is not as the ratio of the density of one medium to the density of another. The second, however, is that these motions are resisted by the medium and are not natural. Yet nobody before him had arrived at these questions, and thus he was more profound than any others." And in the copy [of Ibn Bājja's work] from which we have written, we found a certain page all by itself, and this was written on it. (in Grant 1974, p. 262)

He then explains that the philosopher of Zaragoza's profound argument lies in his conception of speed as mathematical magnitude. Failure to recognise the far-reaching philosophical implication of Ibn Bājja's groundbreaking mathematical approach to motion that paves the way for the mathematisation of falling bodies has led the traditionalist philosopher to accuse his compatriot of conceptual confusion:

> The problem at hand is really sophistical, though difficult, for it is all based upon a confusion of categories. And this is so because, as he assimilates motion to a line, he contends that what occurs apropos a line, occurs apropos motion… The cause of this error lies in the judgment that slowness and speed are motions added to, and subtracted from, a motion in the same manner as a line is added to or subtracted from, a line. (ibid., pp. 259–260)

Argumentation.[9] Islamic jurisprudence was an important source of inspiration for the constitution of future scientific disciplines like Arabic grammar and Algebra, and its new way of arguing has redefined the nature of scientific inquiry since it was so widely adopted that it became a universal method of investigation. It was particularly used by the founder of modern optics and astronomy Ibn al-Haytham (d. 940) in his foundational astronomical work *al-Shukūk* or *Doubts About Ptolemy* to successfully bring change from within the well-established Ptolemaic tradition that was stagnating for centuries.[10]

The study of the Arabic language is the second discipline that played a major role in the universalisation of knowledge. Linguistic studies were developed during the eight-century to such extent that they have made possible the translation of all known scientific works produced by previous civilisations, and one of the main motivations for the codification of Arabic grammar was to ensure the stability and expansion of the Arabic language because of its widespread use by non-native Arab speakers. The process of translations is one of the important means that contributed to getting rid of those metaphysical and cultural concepts that might threaten the universality of knowledge; the Arabic language has in fact acted as some sort of filter through which only scientific thoughts are allowed to pass. The outcome of this process of acquisition is that all knowledge becomes accessible to everyone. The vast production of Arabic scientific literature with technical and rich vocabulary makes Arabic a world language, it was no longer seen as a language of some specific people but belongs to everyone. Because Arabic was the only global language in all walks of life, including science and philosophy, knowledge was promoted to an inter-cultural and inter-national levels, it was no longer linked to a

[9]*Al-jadal* is the *maṣdar* of the relational verb *jādal yujādilu* which means to argue with someone over something hence to counterargue i.e. to come up with an argument that challenges someone's claim. *Al-jadal* of the jurists, whose conceptual foundation is based on epistemic grounds, should not though be confused with Aristotle's dialectic also rendered as *jadal* by the translators of the Greek works (Miller 1985, Chap. II, p. 52). Like *al-qiyās,* i.e. legal analogical reasoning, the jurists conceive their *jadal* as a logic of discovery by which they aim to increase the body of legal knowledge i.e. "to know what is previously unknown معرفة ما لم يعلم" (e.g. al-Juwainī 1979, p. 46). Despite the existence of a vast Arabic literature on *jadal*, scholars have shown little interest in the topic, which is unfortunate given its historical and epistemic importance and relevance to modern argumentation theory. One of the few such studies is Miller's unpublished dissertation *Islamic Disputation Theory: a Study of the Development of Dialectic in Islam from the Tenth Through Fourteenth Centuries*. It is an excellent survey on how the jurists and the *mutakallimūn* have developed the topic from its early beginning as *ādāb al-jadal*, specific to legal reasoning, to its final transformation as *'ilm al-munāẓara* by al-Samarqandī (1250–1310), giving rise to the formation of a new logical science that provides the first logical model of argumentation to be used in scientific debates. It is worth mentioning a forthcoming second work on the topic entitled *The Dialectical Forge: Juridical Disputation and the Evolution of Islamic Law* by W. E. Young to appear in LEUS, which is a landmark on the application of *jadal* to juridical reasoning.

[10]In the first discussion of his *Tahāfut*, al-Ghazālī (1058–1111) explains why refutation according to the *kalām*-type *jadal* matters: "Counterargument necessarily demonstrates the flaw of the proponent's argument, and many aspects of the problem are dissolved in evaluating counterarguments and arguments." (Al-Ghazālī 2002, p. 46 § 134)

specific culture but became the property of all humanity. Historically, the Arabic language showed for the first time the possibility of the construction of a unified corpus of knowledge able to work as a trans-cultural vehicle for the transmission of scientific and philosophical thoughts from one culture to another.

Islamic jurisprudence and linguistics were instrumental in the reinvention of mathematics.[11] The emergence of arithmetic and algebra marks the triumph of the new epistemic attitude that bounds theory to action, it is the emergence of usefulness. Theory and practice are not opposed to each other as in the Greek tradition, they are intrinsically and dynamically linked by usefulness which is regarded as evidence for the soundness of the theory: a theory should improve an existing practice and practice should improve the elaboration of the theory; it is considered as the ultimate test for any discourse since any acquired knowledge must yield sooner or later some tangible results.

These three innovative disciplines, which constitute the basic education of Arabic-Islamic intellectuals, have enriched the scientific and philosophical corpus with new concepts and modes of thought that opened up an immense number of new horizons for their development. Of the three it is undoubtedly the reinvention of mathematics which not only carried much farther than any other discipline the de-hellinisation of the Greek heritage but more importantly brought the whole process to completion as it expanded.[12] Algebra in particular has created a dynamic that transformed mathematics and with it the rest of scientific disciplines including philosophy. Al-Kindī (801–873), a mathematician, seems to have foreseen the implications of the renewal of mathematical studies on ancient philosophy, his insight led him to require that mathematics should be used as an indispensable instrument of analysis and investigation in philosophical activity:

> This is the number of his [Aristotle] books, that we have already mentioned, and which a perfect philosopher needs to know, after mathematics, that is to say, the mathematics I have defined by name. For if somebody is lacking in mathematical knowledge, that is, arithmetic, geometry, astronomy and music, and thereafter uses these books throughout his life, he will not be able to complete his knowledge of them, and all his efforts will allow him only to master the 'ability' to repeat if he can remember by heart. As for their deep knowledge and the way to acquire it, these are absolutely non-existent if he has no knowledge of mathematics (in Rashed 2008, pp. 156–157).

The first Arabic philosopher has in effect established a new philosophical tradition that binds philosophy to mathematics. Rashed further points out that the links between philosophy and mathematics are essential to the reconstitution of al-Kindī's system; his *Philosophy Can Only Be Acquired Through the Mathematical Discipline* is an announcement of a new research programme designed to make mathematics a model for philosophical analysis. Al-Fārābī (872–950) belongs to this new tradition founded by al-Kindī, his mathematical and

[11]About the role of linguistics and the legal sciences in the emergence of algebra, see Rashed's introduction to his translation of al-Khwārizmī's work, Sects. 1.4 and 1.5, al-Khwārizmī 2007.

[12]Some of what historians call scientific revolutions, such as the one that took place in the 17th century, can thus be seen as the most important stages in the de-hellenisation process.

philosophical works, which bring much closer the two disciplines, pave the way for the transformation of philosophy by his successor. We will show indeed that Ibn Sīnā is the architect of a major philosophical change from within the dominating Aristotelian tradition triggered by Abū Naṣr al-Fārābī, the only philosopher that he appreciates and praises among all his predecessors.

The present work is structured into five chapters. The first chapter presents the background of Ibn Sīnā intellectual formation which is essential to understanding what led him to reinvent epistemology. It was his discovery of the nature of the relationship between mathematics and the mind. Chapter 2 is devoted to Ibn Sīnā systematic refutation of the Greek conception of number. Chapter 3 sets out the basics of Ibn Sīnā's theory of knowledge needed for the construction of numbers. We will see how the physician tackles the question from the ontological point of view, intentionality turns out to be crucial in showing the ontological dependence of numbers, and will be even more critical to their construction (Chap. 4). The logico-epistemic construction of numbers is discussed in Chap. 4 in which Ibn Sīnā, thanks to the concept of instantiation, meets a challenge that would be mounted by Frege eight centuries later. By epistemic we mean not only the construction of numbers by the human mind but more importantly how the epistemic agent becomes aware of his own constructions. Our conclusion (Chap. 5) contains some remarks about the new philosophical situation created by Ibn Sīnā's powerful thought. Let us finally emphasise that Ibn Sīnā's enterprise is exclusively motivated by questions of understanding i.e. questions such as: what are numbers? What is their meaning and where are they? The answers which should be both rooted in mathematical practice and should motivate the further development of this practice. From this point of view Ibn Sīnā's inquiry is also intended to be a contribution to explain the actual genesis and evolution of mathematics by the human mind.

Chapter 2
Ibn Sīnā and the Reinvention of Epistemology

2.1 Ibn Sīnā by Himself

Abū ʿAlī al-Ḥusayn Ibn ʿAbd Allāh Ibn Al-Ḥasan ibn ʻAli Ibn Sīnā was born in 980 CE/370 AH in his mother's home village of Afshana near Bukhārā, the capital of the Sāmānid dynasty in central Asia (present day Uzbekistan). He came from a middle class family. His father ʿAbd Allāh was from Balkh, an important town of the Sāmānid empire in what is today Afghanistan, but moved to Bukhārā to serve as a governor of a nearby village. A few years after the birth of his brother, Ibn Sīnā's family settled in the capital Bukhārā where he received his entire education. He died in 1037 CE/428 AH in Hamadhān, Iran. After introducing himself very briefly in his autobiography, Abū ʿAlī spells out the course of his education in the intellectual capital by dividing it into two main stages. The first stage was a basic education consisting of non-Greek or Arabic-Islamic sciences: Qur'ān, Arabic literature, fiqh or Islamic law and jurisprudence including *al-jadal*, arithmetic and algebra.[1]

> A teacher of the Qur'ān and a teacher of literature were provided for me, and when I reached the age of ten I had finished the Qurān and many works of literature, so that people were greatly amazed at me. [...] Then he [my father] sent me to a vegetable seller who used Indian arithmetic, I was thus learning from him. [...] Before his [al-Nātilī] arrival I

---

[1]Ibn Sīnā tells us later that he has mastered the mathematical sciences, which means that he also learned algebra, as evidenced by his mathematical works. As will be explained later, many bio-bibliographers such as al-Bayhaqī are right when they point out:

> When he was ten years old, he knew certain fundamental texts of literature by heart. His father was studying and reflecting upon an opuscule of the Brothers of the Purity. He also reflected over it. His father took him to a greengrocer named Maḥmūd al-Massāḥ who knew Indian calculation and algebra and al-muqābala (Al-Bayhaqī 1946, p. 53).

Ibn al-ʿImād, who quotes Ibn Khallikān, gives this biographical anecdote in similar words: "When he was ten years old, he improved his knowledge in the science of the Glorious Qur'ān, literature, and he knew certain religious foundations by heart, Indian calculation and algebra and al-muqābala" (Ibn al-ʿImād n.d., III, p. 234; see also Ibn Khallikān 1969, II, pp. 157–158).

H. Tahiri, *Mathematics and the Mind*,
SpringerBriefs in Philosophy, DOI 10.1007/978-3-319-25238-4_2

had devoted myself to jurisprudence with frequent visits to Ismāʿīl al-Zāhid about it. I was one of the most agile questioners, having become acquainted with the methods of request and the modes of interception to the respondent (طرق المطالبة و وجوه الإعتراض على المجيب)[2] in

[2]It is particularly interesting that Ibn Sīnā provides these kinds of details in his short autobiography for it shows that he was given advanced training in legal disputation by Ismāʿīl al-Zāhid (d. 1012) who was a ḥanafī jurist. How advanced was it? This is indicated by the expression: the methods of *al-muṭālaba* and the modes of *al-iʿtirāḍ*. Most scholars have struggled to translate *al-muṭālaba* (المطالبة) and *al-iʿtirāḍ* (الاعتراض), and Gohlman rendered them as prosecution and rebuttal, respectively, which is not accurate. The inaccurate translations are due to the fact that the translators seemed to be puzzled by the literal meaning of the words. Literally, *muṭālaba* means asking, but not in the sense of asking questions but of asking for or requesting something and hence is closer to requesting. Furthermore, *iʿtirāḍ* means interception. Literally then, the sentence should be translated as methods of request and modes of interception. But, request what? And interception of what? The literal meaning seems puzzling but, in fact, it is not so; it only sounds incomplete because the two terms do not have the same technical meaning in English as in Arabic. In the *jadal* literature, they are just two from the arsenal of rules and argumentation techniques that can be used by the questioner (*al-saʾil*) to attack the respondent (*al-mujīb*), and their use is governed by the rules of the argumentation procedure. As very specific terms of the new logical science, it is not sufficient to suggest the complete translation of the two concepts without explaining their meaning. By the same token, it will become clear why Ibn Sīnā mentioned just these two technical words. I will very briefly explain the underlying idea of *jadal* in 11th century Islamic jurisprudence, which would have been the kind of *jadal* roughly practiced in Ibn Sīnā's time (for an extensive analysis of the state of *jadal* in this period, see Miller 1985, Chap. III, p. 87). My presentation is based on the discussions of the two prominent jurists Abū al-Walīd al-Bājī's *Kitāb al-minhāj fī tartīb al-ḥijāj* and Imām al-Ḥaramain al-Juwainī's *al-Kāfiya fī al-jadal*, the teacher of al-Ghazālī. The argumentation procedure is comprised of two stages. First stage: understanding the point of the dispute. This is the aim of the following three moves that should be introduced in the following order: (1) What is your opinion? (the whatness of the opinion); (2) What is your evidence?" (the whatness of the evidence); (3) *Muṭālaba bi-tabyīni ad-dalīl* (request for clarification of the evidence). This third move is introduced by al-Juwainī, who defined it as "challenging the opponent to clarify his proof (مؤاخذة الخصم بتبين الحجة)" (al-Juwainī 1979, p. 68), and is of two types: (*i*) request for clarifying the basis (*aṣl*) of proof (*dalāla*) and its establishment and (*ii*) request for clarifying the mode of proof (*wajh ad-dalāla*)." (ibid.). Al-Bājī calls this third move *muṭālaba bit-taṣḥīḥ* or request for verification of the proof (al-Bājī 1978, pp. 40–41), which illuminates al-Juwainī's label since the request for clarification is aimed at verifying that such and such is the case. We choose this way of representing the third move because it gives us a better idea of how to characterize this entire phase, for it turns out that the aim of the three successive questions is to make sure that the opponent understands the actual position of the proponent. While keeping *muṭālaba bi-tabyīni ad-dalīl* (*muṭālaba* can in fact be dropped) for the third move, the three questions can be grouped under the general name *Muṭālaba bit-taṣḥīḥ* or *Muṭālaba bit-taḥqīq* which can then be translated as a Request to establish the facts. Only after making sure that the questioner understands the respondent's thesis and the evidence on which it is based can he begin the process of its refutation. But, as pointed out by al-Juwainī, the questioner can go directly to the refutation stage if the respondent's thesis and the evidence on which it is based are widely known to him (al-Juwainī 1979, pp. 79–80). Second stage: the refutation procedure. The principle of *jadal* is what al-Juwainī calls *al-ilzām bil-muqābala* or the necessity for exchanging arguments, i.e. the respondent is obliged to answer all legal attacks on his evidence. The questioner is then provided with all kinds of techniques that can be used to defeat the evidence of his opponent. *Iʿtirāḍ* is just one of these techniques and is defined by al-Juwainī as follows: "opposing (*muqābala*) the opponent's claim by an argument that prevents him from attaining his goal; and it is said: preventing (*mumānaʿa*) the opponent from proving his point by opposing an argument of equal probative force

> the manner which the practitioners of it [jurisprudence] follow.[3] (Gholman 1974, pp. 19–21)

In the second stage, he was introduced to Greek science and philosophy by another teacher, al-Nātilī, in the following order as mentioned by him—logic beginning with the book of *Isagoge,* the book of Euclid, *The Almagest*—before he went on to study natural sciences (الطبيعيات), metaphysics (الإلهيات) and medicine by himself.

This rapid survey shows that Abū ʿAlī received a much richer interdisciplinary education than that which the Arabic-to-Greek tradition wants us to believe, and the Aristotelian corpus does not constitute the basis of his education as this latter is dominated by the mathematical sciences, mastery of which is unusual in the formation of a traditional philosopher. Is this the case by chance? By no means, as we shall see. What follows represents what seems to be a significant development in Ibn Sīnā's thought, which has had major implications for philosophy. After "mastering the logical, natural and mathematical sciences", we are surprised by the young prodigy's bold admission that he simply could not make sense of what a great classicist calls "the Mount Everest of Aristotle's treatises"[4]:

(Footnote 2 continued)

(*bi-musāwātihi*)". (ibid., p. 67). He justifies his definition by recalling that it is close to its intuitive meaning (لغة): someone intercepts me (*iʿtaraḍanī*) on my way means he prevents me (*manaʿanī*) from following it. After providing a second example, he concludes: "since the opponent prevents the proponent from reaching his goal by opposing an argument with equal probative force, it is called interception in close proximity to the natural language." (ibid.). Literally then, it can be translated as interception of the proof (of the thesis). More generally, *iʿtirāḍ* represents the kind of arguments that attack the rules of inference used by the proponent to establish the conclusion of his claim. In our modern argumentation theory, some formal systems term this kind of attacks as undercutting an inference (Prakken 2002, p. 231). Al-Bājī stresses its importance by describing *iʿtirāḍ* as the argument that attacks "the heart of the evidence (*nafs ad-dalīl*) rendering it void (*bimā yubṭiluhu*)" (al-Bājī 1978, p. 41) and, by this, he means the internal structure of the evidence. He discusses fifteen types of *iʿtirāḍ*, whereas al-Juwainī examines nine. With just two keywords, Ibn Sīnā succeeds in capturing the essence of *jadal* as practiced in the eleventh century. Ibn Sīnā seems to be proud of his courses in Islamic sciences and it appears that he was a formidable opponent. His excellent training in law, jurisprudence and *jadal* was thus very important in his intellectual formation as they contributed to sharpening his philosophical investigations as evidenced in his refutation of the Greeks' conception of number. His refutation of Aristotles' *Metaphysics* (books M and N) in particular confirms that he learned one of the basic lessons of *jadal* of the jurists, i.e. any successful refutation should be preceded by a comprehensive understanding of the opponent's theory. This is just one of many examples, and there are undoubtedly others that show the knowledge of the two powerful traditions, juridical and *kalām*, are essential to understanding the development of his philosophical thought and of the history of Arabic philosophy in general.

[3]The pagination refers to Gohlman's parallel Arabic-English text.

[4]Ross (1995, p. ix).

> I read the *Metaphysics*, but I could not comprehend its content, and the aim of its author [واضعه][5] was confused to me, to the point I reread it forty times and ended up having it memorized. In spite of this I still did not understand it nor what was intended by it, and I despaired of myself and said, "there is no way of understanding this book." (ibid., pp. 32–33)

It looks as if Ibn Sīnā's exceptional correct guessing failed him time and again in attempting to penetrate, let alone recover, Aristotle's most cherished heritage. Why was he unable to understand the content of *Metaphysics* despite mastering the new sciences of his time? Why does *Metaphysics* above all other works prove to be so recalcitrant? And what more does he need to learn to enable him to understand the metaphysics of the ancients? The precocious teenager was probably not aware that he was attacking the founder of the powerful Aristotelian tradition by declaring that not only was he unable to understand the content of one of his best works, but even that "the aim of its author [واضعه] was confused to me".

This was not the first time that Ibn Sīnā experienced difficulties with metaphysics. He tells us that he first heard about it when he was a child:

> My father was one of those who responded to the call of the Egyptians and was considered one of the Ismā'īliyya. From them, he, as well as my brother, heard the account of the soul and the intellect in the special manner in which they speak about it and know it. Sometimes they used to discuss this among themselves while I was listening to them and understanding what they were saying, but my soul would not accept it, and so they began inviting me to respond to it. They were also frequently talking about philosophy, geometry and Indian arithmetic. Moreover he [my father] used to send me to a vegetable seller who used Indian arithmetic, I was thus learning from him. (ibid., pp. 18–21)

This is an extremely important passage as it gives us an idea about the education of his time and the kind of topics discussed in his days that significantly impacted his education. He specifically tells us that he was not satisfied with the doctrines circulating in his milieu about the soul and the intellect. And he further points out that it is in the context of the discussion on the relationship between philosophy and mathematics that his father took special care to provide his brilliant son with some basic mathematical training. But why mathematics first? And how can we explain that he was taught two fundamental mathematical disciplines, unknown to Greek mathematicians and philosophers, by just a greengrocer? Widespread mathematical knowledge was symptomatic of the social revolution brought about by the mathematical revolution. The development of arithmetic and algebra greatly contributed to raising the educational standard of society, which became aware of the necessity of mastering mathematics in order to reach the top of the social hierarchy. Since al-Khwārizmī, algebra was widely taught and used as a practical method of arithmetical problem-solving by civil society including jurists, engineers, secretaries, merchants, surveyors and accountants. As a result, middle class families, and even

[5]Ibn Sīnā had the opportunity to mention Aristotle by name in this passage but did not and, in fact, the latter was not mentioned by his name let alone by his title in his entire autobiography. It is also noteworthy that he sharply rebukes his disciple al-Juzjānī when he "asked him to comment on the works of Aristotle, but he said that he was not free to do so." (ibid., p. 55)

modest ones, were very anxious that their children be taught mathematics, mainly arithmetic and algebra, to ensure a better future for them. A new generation emerged in which knowledge of arithmetic and algebra were part of their culture and the basis of their intellectual formation. The job of the philosopher was consequently transformed, as philosophy could not be practiced as it was because of the emergence of the new sciences and in particular mathematics due to its distinctive epistemic status.[6] Like his predecessors, Ibn Sīnā was part of this new class of philosophers who were first taught the new mathematics before ancient philosophy, and it would only be a matter of time before tensions between the old inherited and newly created conception of knowledge would surface. The education of Ibn Sīnā simply reflects the new scientific and social context that arose in the eighth century from the three innovative disciplines of law, linguistics and mathematics. It is because he mastered the new mathematics that he was able, as he tells us in his autobiography, to occupy important jobs like financial administrative posts. A little-known fact is that he also worked for a while as a jurist:

> I assumed some post in the financial administration of the Sultan. But necessity then led me to abandon Bukhārā and move to Gurgānj, where Abū al-Ḥusayn al-Suhaylī, a lover of the sciences (المحب لهذه العلوم), was a minister. I was presented to the Amīr there, ‘Ali Ibn Ma'mun; at that time I was in lawyer's dress (بزي الفقهاء), with a fold of the mantle under my chin. They gave me a monthly salary which provided enough for someone like me. (ibid., pp. 40–41)

It is this highly educated and prosperous society that identified itself with knowledge which provided Ibn Sīnā with the topic of his future project and enabled him to master all the known sciences. In his case, it is his father, a very religious man, who encouraged his prodigy to learn more[7] about the soul and the intellect and had the remarkable insight that mathematics could be very useful in helping clarify his thought regarding such complex religious philosophical issues. What Ibn Sīnā's father did not know is that he was preparing his beloved son for an ambitious project that would transform philosophy using mathematics. This sounds intriguing in many respects, for how can mathematics clarify strange concepts such as the soul and the intellect? How could such clarification lead to the transformation of philosophy? And what does mathematics have to do with the soul and the intellect in the first place? To his amazement, some of the questions that embroiled the young Ibn Sīnā were already being tackled by one of his predecessors who belonged to the same class of new philosophers.

---

[6]Besides the practical benefit of mathematics, its theoretical usefulness was also recognised and very much appreciated. All the philosophical disciplines, including Aristotelian logic, were subject to bitter controversy among the various traditions. Mathematics, on the other hand, was highly respected by society as a whole and unanimously regarded as a model of rigour, certainty and rationality.

[7]According to one of the best known quranic verses: "Say: My Lord increase me in knowledge و قل رب زدني علما" (20: 114).

2.2 Ibn Sīnā Rediscovers al-Fārābī: The Man Whom He Wishes to Meet!

In his aforementioned significant paper in which he shows how philosophy was shaped by the new mathematics, Rashed establishes the link and the relevance of al-Fārābī's mathematical investigations to those of Ibn Sīnā. However, his presentation seems to be so condensed that most scholars have overlooked his important findings. What follows is an elaboration of his profound insights.

2.2.1 From Metaphysics to Mathematical Knowledge

Al-Fārābī's works were ignored for decades by the Baghdad Peripatetics, who dominated the philosophical scene from the ninth century, until he was rediscovered by chance by Ibn Sīnā who made him famous by immediately recognising his originality.[8] The discovery of al-Fārābī by Ibn Sīnā represents one of the sensational moments in the history of philosophy.

---

[8]To understand what seems to be a curious attempt to transplant the domination of the same old tradition into a different socio-epistemic context as if nothing had changed whatsoever, we need to understand better the context of the translation of Aristotelian works, and the way they were introduced, presented and used, which is as important as the works themselves because of its important consequences. For philosophy to emerge as an independent tradition in a knowledge-based society and survive in the highly competitive intellectual milieu, it needed to rely on a strong figure. The best candidate to serve this purpose was Aristotle because of his historical influence in the Greek tradition, although it seems this alone was not enough. To face the formidable challenges posed by the dynamic of the new cultural environment, the philosophical tradition not only found itself systematically dependent on the thought of the ancient philosopher to the extent that all his known works were translated, it also felt the need to get support from the parallel emergence of a secondary literature which erected the Stagirite to an almost infallible authority. The so-called wisdom (*al-ḥikma*) tradition, which seemed to be initiated by translators of the Greek philosophical works such as the influential Ḥunayn ibn Isḥaq (810–873), appears to have been mainly designed for public consumption since they were written in the form of collections of ethical sayings and anecdotes attributed to past philosophers. It was an attempt to strengthen and enhance the position of the philosophical tradition in the intellectual scene by showing that the ancient Greek philosophers also shared the same values as those of the Arabic Islamic society. Some of these authors aimed to achieve more by claiming that philosophy was completed by what they call the first teacher who should continue to be eternally followed, as eloquently reported by an unknown scribe:

> Aristotle is the first teacher, the seal of the ancient philosophers (*al-ḥukamā'* الحكماء) and the model of the learned people who followed their path. He organised wisdom and established it, he refined philosophy and wrote it down exactly; it is he who put logic at the top and set the foundation for all the rest of sciences. He thus became the medium through which the ancients were to benefit the moderns, and the means by which the successors were to acquire the benefits of the ancestors. No sooner had he content himself to pour upon the latter generation what the former one had caught than he added to every kind of knowledge many times what they had come up with (أضاف إلى كل نوع أضعاف ما أتوا به), bringing it thereby to completion and perfection (فصيره بذلك أتم و أكمل). (ms. Arabe 202, f. 29r)

> But one day in the afternoon when I was at the booksellers' quarter, a salesman approached with a book in his hand which he was calling out for sale. He offered it to me, but I refused it with disgust, believing that there was no benefit in this science. But he said to me, "buy it, because its owner needs the money and so it is cheap. I will sell it to you for three *dirhams.*" So I bought it and it turned out to be Abū Naṣr al-Fārābī's book *Fī Aghrāḍ kitāb mā ba'da aṭ-ṭabī'a* (*On the Aims of the Metaphysics.*) I returned home and was quick to read it, and instantly the aims of that book became clear to me so much so I had it memorized by heart. I rejoiced at this and the next day gave much in alms to the poor in gratitude to God Exalted. (ibid., pp. 32–35)

The Aims of the Metaphysics seems to be the first work of al-Fārābī which Ibn Sīnā read; a four page review in which al-Fārābī confirms the difficulty of Aristotle's work.

> None of the ancient philosophers has properly commented on this book as they have done of his [other] books, but what can be found at most is an incomplete commentary on chapter Lambda by Alexander [of Aphrodisias] and a complete one by Themistius. As for the other chapters, either they have not been commented upon or no commentary has

(Footnote 8 continued)

This passage taken from an Arabic manuscript entitled *Al-Kalimāt al-Mukhtāra min Kalām al-Ḥukamā' al-Arba'a al-Mutaqaddimīn* (preserved in La Bibliothèque Nationale de France and archived as Arabe 202 f.16–36) seems to be a copy of one of the earliest writings in the genre since it presents, as the title indicates, a selection of sayings of only four ancient Greek philosophers: Pythagoras, Socrates, Plato and Aristotle. Besides being the first teacher and the classifier of all sciences, another anecdote, which is currently little known but widely reported at the time in the wisdom literature, identifies Aristotle with Intellect itself (!العقل). As a result philosophy as a complete and closed corpus came to be identified with Aristotle. It seems then that the aim of the project of translating the philosophical works was to perpetuate an old tradition by continuing to generate a series of generations of Aristotelians. This inherited traditional approach to philosophy, which was at odds with the dynamic scientific context in which the translation took place, would, sooner rather than later, inevitably lead to an internal struggle with the rise of a new philosophical movement led by al-Kindī aimed at grounding philosophy in the latest advancements of knowledge. However, the emergence of a generation of new philosophers such as al-Fārābī, who was working in that Aristotelian circle, would find it difficult to openly criticize the first teacher. Even a scholar such as Badawi, who edited Ḥunayn's *Ādāb al-falāsifa* (*The Ethic of the Philosophers*), noted that he had distanced himself from the Aristotelian tradition that tended to make its founder the teacher that could never be surpassed:

> There are some similarities between what al-Fārābī wrote about the school of the philosophers and what Ḥunayn mentioned in the beginning of his book but it seems that al-Fārābī' sources are different from those of Ḥunayn because of the clear differences between their discourse; differences that indicate that al-Fārābī did not base his reports on Ḥunyan. (in Ḥunayn 1985, p. 27)

To challenge the Aristotelians, the philosopher of Bagdad then had to innovate to show them that their teacher was far from having the last word on every topic. His strategy, as we shall see, consisted of expressing his significant non-Aristotelian ideas without making explicit his criticism of Aristotle's views. This was the strategy, followed later by Ibn Sīnā, that would trigger the long but unstoppable process of change.

> survived to our times—since upon examining the books of the later Peripatetics,[9] it may be assumed that Alexander did comment on the entire book.[10] (Al-Fārābī 1890, p. 34)

As a result, Aristotelians reduce metaphysics to theology by ignoring topics they find hard to understand:

> Many people have the preconceived notion that the purport of this book and its content is to discuss the Creator (may He be glorified and exalted), the intellect, the soul, and other related topics, and that the science of metaphysics and the science of the uniqueness of God (علم التوحيد) are one and the same thing. Consequently, we find that most people who study it are perplexed and misguided by it, since we find that most of the talk in it is devoid of any such aim, or rather, we find that the only talk specifically related to this aim is that in the eleventh chapter, that is, the one designated by the letter Lambda.[11] (ibid., p. 34)

Hence his "intention (قصدنا) in this treatise is to indicate the aim that is contained in the book of Aristotle known as *Metaphysics* and the primary divisions (الأقسام الأولى) which it has." (ibid., p. 34). If theology is not the sole subject of *Metaphysics* but only one of its chapters, what are the other chapters about? Answering this question was one of al-Fārābī's aims and the difficult task was to clearly identify the non-theological parts of metaphysics. The approach he adopted was to try to point out in what way the latter differed from two closer prominent theoretical disciplines, i.e. natural philosophy and mathematics. It seems as if the constitution of metaphysics as an independent discipline depends on some kind of metatheory that distinguishes and classifies the various scientific disciplines according to their subject matter. This task is entrusted to what seems to be the powerful metatheoretical discipline that remains implicit in the Aristotelian system, i.e. the so-called classification of the sciences. Through primary divisions, al-Fārābī means to provide the main distinction between the three theoretical disciplines of natural philosophy, mathematics and metaphysics by comparing their subject matter. In doing so, he was led to characterising mathematics in a rather unusual fashion.

> Although mathematics is higher than natural science—since its objects are abstracted from matter—it most certainly should not be called the science of metaphysics because its

[9]In this interesting passage, al-Fārābī shows us that not only was he fully aware of the domination of the Aristotelian tradition since he speaks of "the later Peripatetics" but, more importantly, of its long stagnation since philosophy had made little progress since its founding; a clear indication that he did not consider himself as one of them, as confirmed by Ibn Sīnā:

> As for Abū Naṣr al-Fārābī, we must have a very high opinion of him and he should not be put on the same group of people (و لا يُجْرى مع القوم في ميدان) for he is all but the most excellent of our predecessors. May god facilitate the meeting with him (لعل الله يُسَهِّل الالتقاء معه), so it shall be useful and beneficial. (in Badawi 1978, p. 122).

[10]لا يوجد للقدماء كلام في شرح هذا الكتاب على وجهه كما هو لسائر الكتب بل إن وجد فلمقالة اللام للإسكندر غير تام و لثامسطيوس تاما. و أما المقالات الأخر فإما لم تشرح و إما ان لم تبق إلى زماننا على أنه يظن إذا نظر في كتب المتأخرين من المشائين أن الإسكندر كان قد فسر الكتاب على التمام.

[11]The pagination refers to the Arabic text edited by F. Dieterici in *al-Fārābī's Philosophische Abhandlungen*, 1890.

objects are abstracted from matter only by *[human] imagination*, not in existence.[12] (ibid., p. 36)

The emphasised sentence exemplifies the originality of al-Fārābī for in it he characterises the specificity of mathematics in a way which cannot be found in Aristotle's *Metaphysics*. In his effort to distinguish the subject matter of metaphysics from that of mathematics, al-Fārābī argues that while the objects of both disciplines are abstracted entities, mathematical objects are abstracted from sensible objects as a result of a specific operation of the mind. By making mathematics a product of the mind, al-Fārābī's analysis illuminates Ibn Sīnā's thought on the epistemic role of the soul in the production of knowledge. This distinction between metaphysics and mathematics had far-reaching implications for it motivated the philosopher of Bagdad to provide his own revolutionary classification of the sciences.[13]

2.2.2 *Algebra: The New Universal Science*

In *al-Aghrāḍ*, al-Fārābī demonstrates that he perfectly grasped Aristotle's *Metaphysics*, to the extent that one would expect him to have followed the latter's classification only to see him instead abandoning it altogether in his *Iḥsā' al-'ulūm* or *The Enumeration of the Sciences*, a book in which he drastically overhauls the classification of the sciences.[14] He ignores Aristotle's two major principles of classification. The first is the sharp theoretical/practical dichotomy which seems to be modelled on the value-based Greek social system in which purely theoretical studies are highly valued at the expense of practical arts which are viewed with disdain.[15] Fully aware of the new society which rather divides itself into

[12] و العلم التعاليمي و إن كان أعلى من علم الطبيعة إذ كانت موضوعاته متجردة عن المواد فليس ينبغي أن يسمّى علم ما بعد الطبيعة لأن تجرد موضوعاته عن المواد وهمي لاوجودي.

[13] Al-Fārābī of course also expressed his original ideas elsewhere other than in his *Aghrāḍ* and Ibn Sīnā certainly read whatever of al-Fārābī's work he could find once he had discovered his first work. To show precisely that it was the new mathematics that led to the discovery of a new philosophical approach, we only present his views about some important questions closely connected to *al-Aghrāḍ*, leaving for future research a systematic study of his theory of knowledge and the extent of its influence on Ibn Sīnā.

[14] It seems so revolutionary that posterity has added little to it as pointed out by one of his successors Sa'īd ibn Aḥmad al-Qurṭubī (d. 1070): "He [al-Fārābī] composed a noble work [the *Iḥsā' al-'ulūm*] in which he enumerated the sciences and indicated the object of each; this treatise, the like of which had never before been composed and the plan of which had never been adopted by any other author, is an indispensable guide to students in the sciences." (in Farmer 1932, p. 561).

[15] Greek society was a slave-based society. The small wealthy intellectual elite did not work and did not need to work and so that they could devote their time to leisure in gymnasiums and museums like the Academy and the Lyceum. Aristotle famously justifies the established social class system by speculating that it was necessary for the emergence of the theoretical sciences such

khaṣṣa or elite and *'amma* or common class, it is not surprising that al-Fārābī adopted criteria forged by the new epistemic context of his time which did not separate theory from action through a new understanding of usefulness. Due to the strong social demand for knowledge, education became a profitable market, attracting many people to the teaching sector. And to help parents provide the best education for their children and identify the pseudo-experts in their subject domain, al-Fārābī established a modern detailed programme according to the most basic knowledge that should be acquired by any student.

> What is contained in this book will be useful to the man if he wants to learn some science among these sciences and examines it, he will know what he will undertake, what he will examine, in what way his examination will benefit him, what benefit it has and by what virtue it will be attained, so that he will be able to engage in what he will undertake in the sciences with knowledge and insight and not blindness and deception. With this book, a person can compare the sciences, he will then know which of them is the best, the most useful; the most perfect, reliable and powerful, and which of them the feeblest, the frailest and the weakest. It will also be useful in finding out he who claims he is familiar with some science among these sciences while it is not the case.[16] (Al-Fārābī 1996, p. 16)

He then classified the sciences into five chapters[17]:

(Footnote 15 continued)

as geometry: "The mathematical arts first took shape in Egypt for there the priestly caste was privileged to have leisure" (*Metaphysics* A. 1. 981b). In stark contrast, the development of science in the Arabic tradition was based on a different model as it was financed by public-private mixed funding. Scientists of the famous institution *bayt al-ḥikma* for example were financed by the political body, other scientists such as astronomers were financed by *al-waqf*, an Islamic-specific mode of financing. The third class of intellectuals were able to live from their own works due to the strong demand for knowledge from a prosperous and well-educated society. Ibn Sīnā, who came from a modest family, was one of them and, as he tells us in his autobiography, his career began at an early age (21 years old) when he began composing a wide variety of books commissioned by middle class people who were willing to pay to get the latest advancements in knowledge before joining the top elite. This socio-epistemic model ensured two kinds of interaction: between the intellectuals themselves who belonged to various traditions, and between the intellectual elite and the rest of society. This is an entirely different social and cultural setting than what we find in Ancient Greece.

[16] وينتفع بما في هذا الكتاب الانسان إذا أراد أن يتعلم علما من هذه العلوم وينظر فيه، علم على ماذا يقدم، و في ماذا ينظر، وأي شيء سيفيده نظره، وما غناء ذلك، وأية فضيلة تنال به، ليكون إقدامه على ما يقدم عليه من العلوم على معرفة وبصيرة، لا على عمى و غرر. وبهذا الكتاب يقدر الإنسان على أن يقيس بين العلوم ، فيعلم أيها الأفضل ، وأيها أنفع ، وأيها تقن وأوثق وأقوى ، وأيها أوهن و أوهى و أضعف. و ينتفع به أيضا في تكشف من ادعى البصر بعلم من هذه العلوم ولم يكن كذلك.

[17] This is the new programme that al-Fārābī proposes to reform philosophical studies. Here is not the place to discuss the modernity of his classification. Sufficient for our purpose is to make some important remarks that are symptomatic of the reinvention of knowledge brought about by the surge in social disciplines since the eighth century. The first is the representation of three powerful traditions of his time, i.e. the linguistics, legal and *kalām* traditions. Logic is separated from natural philosophy and metaphysics and placed at the intersection between linguistics and mathematics. It is particularly interesting to note that the new philosopher relegates the two main traditional philosophical disciplines of natural philosophy and metaphysics to fourth place while linguistics comes first. At least two reasons that he simply could not ignore explain this decision, which confirms once again that the philosopher was a thinker of his time. His detailed treatment shows how far linguistics had developed since the eighth century to become one of the most advanced

The first chapter on the science of language and its parts (في علم اللسان و أجزائه);
The second chapter on the science of logic and its parts (في علم المنطق و أجزائه);
The third chapter on the science of mathematics and its parts, comprising the science of number, geometry, optics and astronomy, the science of music, the science of weights and the science of “ingenious devices” (to this we shall return later);
(في علوم التعاليم و أجزائه و هي: العدد، والهندسة، وعلم المناظر، و علم النجوم التعليمي، وعلم الموسيقى، وعلم الأثقال، و علوم الحيل)
The fourth chapter on the science of nature and its parts and on theology and its parts;
(في العلم الطبيعي و أجزائه، و في العلم الإلهي و أجزائه)
The fifth chapter on the science of politics and its parts, on the science of jurisprudence, and the science of *kalām* (في العلم المدني وأجزائه، و في علم الفقه، و علم الكلام).

As for the classification of the theoretical disciplines, al-Fārābī ignores Aristotelian criteria of motion and separation from matter. We will examine his classification of just two of the mathematical sciences he enumerated: the science of number and the science of ingenious devices. He tells us that the science of number is made up of two branches: the science of practical number and the science of theoretical number and begins by defining them as follows:

> The science of practical number examines the numbers inasmuch as they are accountable numbers that needs to determine the number of bodies and similar things such as men, horses, dinars, dirhams or other things which are accountable. They are the numbers that people use in merchant transactions and civil life.[18]
>
> As for the science of theoretical number, it examines the numbers *absolutely* as abstracted from the bodies *in the mind* (مجردة في الذهن), and from *all* what is counted. It examines them *purified* (ملخصة) from all what could be counted with by sensible things. And this is the science which is involved in *all* the sciences (جملة العلوم).[19] (ibid. p. 50, my emphasis)

(Footnote 17 continued)

scientific disciplines as shown by the sheer number of volumes on the Arabic language (series of monographs, dictionaries and encyclopedias) produced by grammarians, lexicographers, litterateurs, etc. Even today, research on the Arabic language is one of the most dynamic and fascinating lines of research on the Arabic tradition. Linguistics not only developed before philosophy but, more importantly, it has greatly influenced the latter by making philosophers aware of the importance and relevance of language analysis in philosophical studies; a fact that was fully grasped by al-Fārābī in the composition of his books such as *Kitāb al-ḥurūf* which, in fact, should rather be called the book of meaning since meaning (الدلالة), an important juridical notion, is its central topic. Linguistics also offers a rich vocabulary that unifies all kinds of knowledge, making it de facto the first lingua franca for centuries to come.

18 فالعملي يفحص عن الاعداد من حيث هى أعداد معدودات تحتاج إلى أن يضبط عددها من أجسام و غيرها، مثل الرجال أو أفراس أو دنانير أو دراهم أو غير ذلك من الأشياء ذوات العدد، و هي التي يتعطاها الجمهور في المعاملات السوقية و المعاملات المدنية.

19 وأما النظري فانه يفحص عن الاعداد بإطلاق على أنها مجردة في الذهن من الأجسام ، وعن كل معدود منها، وإنما ينظر فيها ملخصة عن كل ما يمكن أن تعد بها من ا لمحسوسات، ومن جهة ما يعم جميع الأعداد التي هي أعداد محسوسات؛ وهذا هو الذي يدخل في جملة العلوم.

For al-Fārābī, whose appreciation of arithmetic is reminiscent of that of the nineteenth century German mathematician Gauss (1777–1855), pure number theory is queen of the sciences. In his next paragraph, he explains that the science of theoretical number involves studying numbers as a structure by examining their properties, such as the property of being even or odd, their relationships and the various operations that can be performed on them. In short, what al-Fārābī calls the science of theoretical number is simply number theory.[20] This distinction is of paramount importance for two reasons: firstly, in itself because of its modernity and, secondly,

[20]Al-Fārābī does not put arithmetic and geometry on the same epistemic level since his description of geometric objects is substantially different from the one he previously provides for the natural numbers.

> Theoretical geometry examines lines and planes of bodies absolutely and generally and in a general manner that subsumes the planes of all bodies (على وجه يعم جميع الأجسام). For the geometrician represents in his mind (يصور في نفسه) lines in the general manner regardless (لا يبالي) of whatever body in which they are; and he represents in his mind planes, being square (التربيع), circularity (التدوير), being triangular (التثليث) in the most general manner regardless of whatever body in which they are (بالوجه الأعم الذي لا يبالي في أي جسم كان); and he represents the physical objects in the most general manner regardless of whatever body in which they are and of whatever matter and sensible thing in which they are. He rather represents them absolutely without making reside in his mind (يقيم في نفسه) a physical body be it a wood, a physical body be it a wall or a physical body be it an iron, but the general physical body of these. (ibid., pp. 51–52)

In this paragraph, al-Fārābī does not simply try to identify theoretical geometry as the study of shapes for example, he is rather aiming to determine its epistemic status. Two points of great importance are discussed. The first: in what way geometry is theoretical? His answer is that geometry is theoretical by the nature of its objects: geometric concepts are general, but what does it mean for a concept to be general? Unlike theoretical arithmetic whose objects have little to do with experience, al-Fārābī is struggling here to determine the relationship between particular objects and the concept to which they give rise. And it is in the first sentence where he captures how a concept is related to its particular objects: the concept is general in the sense that it is represented in the mind of the epistemic subject as subsuming all particular objects. And he goes further by spelling out what this specific mental representation amounts to. In the last sentence where he speaks of the mind as a location ("reside in the mind") for the general concepts, al-Fārābī suggests that the mind is an ontological region in which its objects can thus be regarded as having a mental existence. This new understanding of ontology paves the way for a major transformation of ancient metaphysics by making a simple presence in the mind a new form of existence and a mental representation an act of creation. In the second point, he describes the cognitive process involved in the formation of the general concept. Regarding the creation of geometric objects, he explains that the mental act that generates the general concept consists in the identification of an invariant in the perception of physical objects. He uses the concept of "disregard" to indicate that that specific mental act which perceives the invariant property inherently involves reference to an external reality. Because of its shape, the concept of triangle for example refers to all physical objects which have the property of being triangle; and by extension geometric constructions point to some extramental objects leaving the question of their realisation, i.e. their actual construction, to practical geometry. Though geometry is theoretical since its objects are general concepts, its subject matter, in which shape is an integral part, is what makes it essentially linked to objects of experience; that is what al-Fārābī is trying to say: unlike number theory which is a pure science, geometry is an empirical science.

because of its historical significance. As will be shown in the next chapter, Aristotle finds it hard to clarify the nature of mathematics since he seems to confuse two important conceptual terms by calling theoretical what is empirical. As an alternative to the underlying conception of the Pythagorean mathematical practice,[21] the author of *Metaphysics* suggests that their manipulation of numbers assumes a reference to some concept serving as a counting unit and on which their existence is dependent: "one means a measure of some plurality, and number means a measured plurality and a plurality of measures" (*Metaphy.*, N. 1088a), which implies that one is not in itself the substance of anything but rather of a particular and definite substance or, as he puts it, "a number, whatever it is, is always a number of something" (ibid., N. 1092b). Al-Fārābī disagrees and refutes Aristotle's claim by arguing that numbers are objects of the mind and, as a result, they do not need to refer to an implicit counting unit since number theory examines numbers themselves as unities and not made of unities. But how does he deal with the question of their existence? For an answer, we have to go to his *Kitāb al-ḥurūf* where he discusses the ontological issue:

> The thing can be said of every thing that has a quiddity, whether it is external to the soul or [merely] conceived of in any way… Whereas the existent is always said of every thing that has a quiddity, external to the soul, and cannot be said of a quiddity merely conceived of. For this reason the thing is more general than the existent.[22] (Al-Fārābī 1970, p. 128).

Al-Fārābī is thus fully aware of many of the problems of *Metaphysics*, one of which is the enormous struggle faced by the doctrine of its author to account for the extant mathematical ontology that was further extended by al-Khwārizmī's algebra. As al-Fārābī pointed out in *The Aim*, according to Aristotle, metaphysics is a universal science since its subject is 'being' in general. But al-Fārābī attacks the old metaphysics by surprisingly arguing that the thing is more general than the existent. It looks as if what can be called the science of the thing is more universal than the

---

[21]Aristotle launched a strong attack on the Pythagoreans by emphasising that they "use stranger principles than those of the natural philosophers, because they take them from the non-sensible, unchangeable world of mathematics. Yet all their discussions are about nature" (*Metaphys.* A 989b)—an odd philosophical approach that he attacks further when it encroaches on his poaching territory, natural philosophy, by claiming that numbers are the cause of things:

> Number, then, whether it be number in general or number composed of units, is neither the cause as agent, nor as matter, nor as ratio and form of things. Nor, of course, is it the final cause. (ibid., N. 1092b)

This strong hostility against the Pythagoreans has further implications for the Aristotelian epistemic status of mathematics. Aristotle showed us that his theory of causes can defeat the Pythagoreans bizarre doctrine, but the author of *Metaphysics* believes that he has established more in terms of the limited relevance of mathematics to philosophy by conceding elsewhere that at best astronomy is "the nearest to philosophy of all the mathematical sciences, since it studies substance which is eternal, whereas the others are concerned with no kind of substance, e.g. the sciences of arithmetic and geometry." (ibid., Λ. 1073b).

[22]والشيء قد يقال على كلّ ما له ماهيّة مّا كيف كان، كان خارج النفس أو كان متصوَّرا على أيّ جهة كان... فإنّ الموجود إنّما يقال على ما له ماهيّة خارج النفس ولا يقال على ما له ماهيّة متصوَّرة فقط، فبهذا يكون الشيء أعمّ من الموجود.

science of being. Is there a science whose object is the thing? A traditional philosopher whose knowledge is limited to the Greek scientific and philosophical output would be puzzled to hear that there is such a thing as the science of the thing; he would be lost if he tried to find it in al-Fārābī's classification of the sciences as he would not be able to recognise that the author of *Iḥsā' al-'ulūm* had put it under a new category called "the science of ingenious devices"

> In arithmetic, the ingenious devices involve, among other things, the science known by our contemporaries under the name of algebra and al-muqābala, and what is similar to it. But this science is common both to arithmetic and geometry.[23] (ibid., pp. 63–64).

This is just another example that shows that philosophy can make little progress if it ignores major scientific achievements. The emergence of new mathematical branches like arithmetic, algebra, the science of ingenious devices and the surge in major social disciplines like law and linguistics have made obsolete the old Aristotelian classification of the sciences.[24] He further explains how algebra comes to wreck the Greek conceptions of arithmetic and geometry by giving rise to a new unifying mathematical concept, i.e. the algebraic quantity, the thing.

> It includes the ingenious devices to determine the numbers that we try to determine and use, those which are rational and irrational the principles of which are given in Euclid's *al-Usṭuqusāt* 10th book, and those which are not mentioned by Euclid. Since the relation of rational to irrational numbers — to one another — is like the relation of numbers to numbers, each number is thus homologous with a certain rational or irrational magnitude. If we determine the numbers which are homologous with magnitude ratios, we then determine these magnitudes in a certain manner. That is why we postulate certain rational numbers to be homologous with rational magnitudes, and certain irrational numbers to be homologous with irrational magnitudes. (ibid., p. 64)

This text is of great historical importance for it demonstrates that al-Fārābī not only was aware of the latest developments in the mathematics of his time, he had also fully grasped the far-reaching philosophical implications of the reinvention of mathematics. It was al-Khwārizmī who created a new formal ontology[25] by making the "thing" such a powerful mathematical concept to refer to both rational and irrational numbers, as Rashed explains:

---

[23] فمنها [علوم الحيل] الحيل العددية، و هي على وجوه كثيرة: منها العلم المعروف عند أهل زماننا بالجبر والمقابلة و ما شاكل ذلك. على أن هذا العلم مشرك للعدد والهندسة.

[24] Al-Khwārizmī's algebra undermines one of the major Aristotelian metaphysical assumptions underlying his classification of the sciences that can be called the incommensurability of kinds, i.e. the impossibility of passing from one kind to another. Like living species, Aristotle has denied the possibility of interaction between not only the various sciences, like mathematics and physics, but also within a given scientific discipline like mathematics, confidently concluding: "Therefore it is impossible to prove a fact by transition from another genus, e.g. a geometrical fact by arithmetic. [...] The axioms may be the same; but where the genus is different, as of arithmetic and geometry, the arithmetical proof cannot be applied to prove the attributes of spatial magnitudes." (*An. Post.*, 75a38).

[25] See also al-Khwārizmī (2007, p. 48).

> If algebra is in fact common to arithmetic and geometry, without in any way giving up its status as science, it is because its very object, the "algebraic unknown", that is, the "thing (الشيء, res)" can refer indifferently to a number or to a geometric magnitude. More than that: since a number can also be irrational, "the thing" designates then a quantity which can be known only by approximation. Accordingly the algebraists' subject matter must be general enough to receive a wide range of contents; but it must moreover exist independently of its own determinations, so that it can always be possible to improve the approximation. The Aristotelian theory is obviously unable to account for the ontological status of such an object. So a new ontology has to be made to intervene that allows us to speak of an object devoid of the character which would none the less enable us to discern what it is the abstraction of; an ontology which must also enable us to know an object without being able to represent it exactly. This is precisely what has been developing in Islamic philosophy since al-Fārābī: an ontology which is "formal" enough, in a way, to meet the requirements mentioned above. (Rashed 2008, pp. 168–169)

We have presented just some views of al-Fārābī's philosophy of mathematics, which represent a major breakthrough that few were capable of recognising at his time. The problem that gave so much trouble to Ibn Sīnā on the nature of the relationship between the soul and mathematics was clarified by his predecessor, i.e. that mathematics is created by the human mind. The significance of al-Fārābī's original ideas is that it opened a new field of research the exploration of which would lead his successor to reinvent epistemology. It is remarkable that since childhood, one specific entity intrigued Ibn Sīnā's thought and dominated his investigations, which distinguished his project from those of all his predecessors; namely, the soul and its relationship to the world. And to the question: What is the soul? He does not try to answer it in the old fashioned way by writing an abstract treatise on the subject. Instead, he relates it to its production by considering the soul as a powerful creation of knowledge—a new topic that generates some unprecedented epistemic questions such as: what is knowledge?; how is it created by the soul?; what is possible for it to know?; how does it know that it knows?; and, more importantly, how can it know more than what it knows? He then used mathematics, the knowledge par excellence, as an instrument of investigation and analysis of the mind by characterising the various mental operations that could explain the production of mathematical knowledge and contribute to its progress through various levels of understanding.

By calling him the second teacher, Aristotelians seem to be proud of al-Fārābī's contributions to the understanding of the Aristotelian doctrine, an admission that, before him, philosophy was in disarray not to say in crisis. The irony is that they did not realise that he was in fact leading a major change in philosophical tradition. This is the subtle point that Ibn Sīnā grasped once he read the first book that fell into his hands, which happened to be *The Aim of the Metaphysics*.

> The purpose of its author [واضعه] was confused to me… it turned out to be Abū Naṣr al-Fārābī's book *Fī Aghrāḍ kitāb mā ba'da aṭ-ṭabī'a*. I returned home and was quick to read it, and instantly the aims of that book became clear to me. (Gohlman 1974, pp. 33–35)

This is the passage that makes al-Fārābī famous and by which it became known that his *al-Aghrāḍ* helped his successor understand Aristotle's *Metaphysics*. But what is little recognised is the underlying major conceptual change that was communicated.

Aristotle's treatise becomes clear to Ibn Sīnā only after reading al-Fārābī's work, whereas beforehand as he admitted he was confused. It turns out that what al-Fārābī's work made him understand was why he was confused, helping him to understand issues that the Stagirite had failed to tackle let alone clarify, i.e. how mathematical knowledge is created by the epistemic subject. In other words, after reading al-Fārābī's *Aghrāḍ*, Ibn Sīnā was able to identify some of the problems that make *Metaphysics* the Mount Everest of Aristotle's treatises, its refutation was just around the corner. Al-Fārābī had already started the process of breaking down the old metaphysics by highlighting some of the weaknesses of Aristotle's ontology, especially regarding mathematics. In *al-Ilāhiyāt* or the new metaphysics, Ibn Sīnā finishes the job by being considerably more precise in his attacks. As a result, the differences within the philosophical tradition became much more serious than had ever been the case and the gap between representatives of the stagnating Aristotelian tradition in particular and Ibn Sīnā's increasingly non-Aristotelian tendency became much wider. Since he belonged to the philosophical circle dominated by the traditional philosophers, he gave them the benefit of the doubt by initially trying to accommodate them. But the ever-increasing gap only made matters worse, and ultimately it became so wide that he could not even have a basic meaningful discussion with them. This persistent dialogue with the deaf made him realise that the old generation were so deeply entrenched in the Aristotelian tradition that they would never change as he was hoping. He learned from his experience that, like his predecessor, his thoughts were so revolutionary that they would only be understood by the generations to come. It is for them that he wrote his *Logic of the Orientals* in which he takes the landmark historical decision to publicly announce the beginning of a new era.

Chapter 3
Refutation of the Greek Conception of Number

Is there in the Arabic tradition room for a conception of the natural numbers that differs, say, from the one of the Greek legacy? And if so, what could it be? What is little known is that questions about the foundations of mathematics which have generated so many lively discussions in the development of contemporary mathematics and logic have equally been a subject of research by the ancients who acknowledged it both as important and difficult. This work can be considered as a recovery of an old new epistemology designed to account for the generation of numbers based on the actual practice of mathematics and conceived for both the understanding and the further development of this practice rather than to corroborate preconceived metaphysical assumptions. Unfortunately, in most of the contemporary analytic philosophy of mathematics the historical roots of this approach have been mostly overlooked.[1] Mathematicians, logicians and historians of mathematics would be surprised to find that what seems to be a central topic of the 19th century philosophy of science was a topic that reinvented epistemology ten centuries earlier. This gap in the understanding of the development of the history of mathematics is by no means an isolated example for we have already drawn the attention of the scientific community to similar cases in mathematics, logic, astronomy and philosophy. Moreover this issue seems to represent another example of the mainstream misconception of the development of science in general and of the role of the Arabic tradition in particular. Indeed, in his monumental work *al-Shifā'* (or *The Healing*), Ibn Sīnā has conducted an unprecedented investigation into the concept of number whose status was left in disarray since the discovery of the irrationals and motivated our author to explore new paths despite the difficulty of the task as he admits:

[1]This chapter is an overview of Ibn Sīnā's theory of knowledge designed to be further examined and discussed in future researches. I by no means want to give the impression that his theory is perfect or answers all the questions and need not be understood as such. The present work should be considered as an outline of Ibn Sīnā's conception of number for it contains interesting answers to some complex questions and as a result should be put under further scrutiny in subsequent researches.

H. Tahiri, *Mathematics and the Mind*,
SpringerBriefs in Philosophy, DOI 10.1007/978-3-319-25238-4_3

How difficult it is for us to say something that is reliable on this topic [of the nature of number].

"فما أعسر علينا أن نقول في هذا الباب شيئا يعتد به" (p. 80, § 3).[2]

The Greeks proved their great ingenuity by demonstrating the existence of irrational numbers, but their solution to the problem was purely and simply to unanimously reject them, as soon as they were discovered, and make arithmetic part of geometry loosing thus its independence status. It is in this context that one can realise the epistemic revolution that took place when in the early 9th century Muḥammad ibn Mūsā al-Khwārizmī tackled the discovery that seems to have shocked the Greeks. In his *Kitāb al-Jabr wa-al-Muqābala*,[3] he showed, among other things, that the irrational numbers obey the same arithmetical rules as the rational numbers, extending thereby the concept of number. Ibn Sīnā was the product of this new world created by al-Khwārizmī's *Algebra* that led him to radically transform the philosophy of mathematics.

The present work is in line with similar researches that have recently focused on Ibn Sīnā's mathematical works conducted by world leading mathematicians and logicians. Roshdi Rashed was the first to point out the importance of mathematics to Ibn Sīnā's philosophy. In his aforementioned paper, Rashed explains how Ibn Sīnā's *al-Arithmāṭīqī* radically shakes up the ancient classification of the four mathematical disciplines with the aim of making them independent of natural philosophy: "from now on, he concludes, all the ontological and cosmological considerations which *burdened* the notion of number are de facto banned from *al-Arithmāṭīqī*, considered thus as a science" (Rashed 2008, p. 166). The independence of mathematics, which was the result of being liberated from natural philosophy that hampered its development for centuries, enables it to reinvent itself and to create the dynamic needed for its sustainable development by studying its own fundamental concepts and methods, and which was recognised as pure mathematics (الرياضيات المحضة). However, according to our view, Ibn Sīnā's epistemic approach searches for epistemic concepts that should rather help the development of mathematics by itself by intrinsically understanding its own structure. His systematic refutation of the Greek conception of number already indicates his originality and illustrates the nature of his logico-epistemic investigation. In Chap. 2 of Book 7 of *al-Ilāhiyāt*, Ibn Sīnā begins by presenting very

[2]Unless indicated otherwise, all Ibn Sīnā's quotations are taken from his *al-Ilāhiyāt*, the number of pages and paragraphs refers to the bilingual edition of the book which was translated by Marmura under the title "The Metaphysics of *The Healing*".

[3]The book is now available thanks to the enormous work of the distinguished historian of mathematics Roshdi Rashed who produced the first Arabic critical edition of al-Khwārizmī's *Algebra* and in parallel a French translation, the first rigorous translation into a European language, with a full commentary on the text. The translation of the book published under the title *Al-Khwārizmī. Le commencement de l'algèbre* (Blanchard-Paris 2007) is preceded by an extensive introduction that puts the landmark founding work of the mathematician of Bagdad into its historical perspective.

briefly his view on the development of philosophy in the Greek period by stressing that it was rhetorical and dominated by natural philosophy:

> We say: every art has a genesis wherein it is raw and unripe, except that after a while it matures and after some more time, it develops and is perfected. For this reason, philosophy in the early period of the Greek's occupation with it was rhetorical. It then became mixed with error and dialectical arguments. Of its divisions, it was the natural which first attracted the masses. They then began to give attention to the mathematics, then to the metaphysical. They were involved in transitions from one part to another that were not sound.[4] (p. 243, § 2)

And why were the Greek doctrines of the three disciplines (physics, mathematics and metaphysics) not sound? His swift answer: "when they first made the transition from what is apprehended by the senses to what is apprehended by the mind, they became confused (و أول ما انتقلوا من المحسوس إلى المعقول تشوشوا)."[5] This criticism seems to particularly target the Greek notion of number. In this context, Ibn Sīnā reviews the various doctrines of the Greeks regarding the existence of numbers and differentiates two main groups: the first one claimed that numbers have separate existence like Plato while the second is represented by two factions: on the one hand Pythagoras and his followers who adopted the idea that objects are constituted by numbers and on the other hand Aristotle, and his disciples, who thought numbers are potentially *in re*. According to Ibn Sīnā, both groups ontologically confused numbers with real objects.

3.1 Criticism of Plato's Realism

Ibn Sīnā explicitly mentions the views of "Plato and his teacher Socrates" on mathematics criticising them for going "into excess in upholding this view, saying that there belongs to humanity one existing in which individuals participate and which continues to exist with their ceasing to exist. This they held was not the sensible, multiple and corruptible meaning and is therefore the intelligible, separable meaning"[6] (p. 244, § 4). From this subordination of the sensible to the intelligible world, Ibn Sīnā concludes "then the intelligible mathematical objects would be things other than those we imagine and intellectually apprehend."[7] Plato

[4]و نقول: إن كل صناعة فإن لها نشأة تكون فيها نيئة فجة غير أنها تنضج بعد حين ثم إنها تزداد و تكمل بعد حين آخر؛ و لذلك كانت الفلسفة في قديم ما اشتغل بها اليونانيون خطبية، ثم خالطها غلط و جدل. و كان السابق إلى الجمهور من أقسامها هو القسم الطبيعي، ثم أخذوا ينتبهون للتعليمي، ثم للآلهي. و كانت لهم انتقالات من بعضها إلى بعض غير سديدة.

[5]This overwhelming attack is one of the strongest signals by which Ibn Sīnā indicates that philosophy as understood and practised by the Greeks is in need for a radical change, he tells us why in this section and shows how in the following.

[6]و كان من المعروف بأفلاطون و معلمه سقراط يفرطان في هذا الرأي و يقولان إن للإنسانية معنى واحدا موجودا يشترك فيه الأشخاص و يبقى مع بطلانها، و ليس هو المعنى المحسوس المتكثر الفاسد فهو إذن المعنى المعقول المفارق.

[7]Ibn Sina's argument strikingly resembles the one that of Ibn al-Haytham advanced in his *al-Shukūk* or *Doubts about Ptolemy*. In his response to Ptolemy's claim that his *Almagest's* complex

seems to have succeeded in making mathematical objects ideal entities, albeit intermediary, by cutting off the intelligible from the sensible; but, according to our author, the Platonists must prove both that the mathematical entities exist and that they exist separately:

> We would then require a *new proof* to establish the existence of the mathematical objects and, after this, to engage in examining the state of their separateness. Thus, what they have done in rendering mathematical objects eternal so as to dispense with *establishing* their existence, and in preoccupying themselves with giving priority to the task of showing their separateness, represents an unreliable course.[8] (pp. 249–250, § 2; my emphasis)

In fact, Ibn Sīnā requires an epistemic argument since he is not satisfied with a metaphysical assumption of existence. This kind of criticism already signals the main motivation of Ibn Sīnā's enquiry: an epistemological elucidation of the concept of number that as such will be fruitful for the development of pure and applied mathematics itself. Let us see how this feature is set forth in Ibn Sīnā's criticism of the second group.

3.2 Criticism of the Pythagoreans' Empiricism

Ibn Sīnā discusses the various doctrines of the Pythagoreans for whom numbers are the principles of everything, and he points out that "some people made these [numbers] principles but did not make them separate. These are the followers of Pythagoras. They composed everything from unity and duality. They made unity within the bound of the good and what is restricted and made duality within the bound of evil and what is unrestricted"[9] (p. 245, § 8). Ibn Sīnā develops several objections to the Pythagorean doctrine, we will discuss only his refutation of their conception of arithmetic. According to Lasserre, the Pythagoreans seem to base their theory of numbers on an epistemic process by means of which "knowing numbers" results from the construction of appropriate figures composed of points, considered as units:

> The mathematicians' attachment to the word unity comes from the fact that they thought of the number as a figure composed of dots, or if we prefer, as a quantity of dots arranged as a

(Footnote 7 continued)

machinery is an imagined planetary theory since it is the combination of purely fictional entities, Ibn al-Haytham concludes that the resulting motion of the various planets occurs solely in his own imagination and as a result has little to do with the real motion of the heavenly bodies: "that [Ptolemy's] configuration produces in his own imagination the motions that belong to the planets "تلك الهيئة تؤدي حركات الكواكب في تخيله على ما هي عليه" (Ibn al-Haytham 1971, p. 38).

[8] فإن كانت [مباينة] له فتكون التعليميات المعقولة أمورا غير التي نتخيلها و نعقلها و نحتاج في إثباتها إلى دليل مستأنف، ثم نشتغل بالنظر في حال مفارقتها. فلا يكون ما عملوا عليه من الإخلاد إلى الإستغناء عن إثباتها و الإنشغال بتقديم الشغل في بيان مفارقتها عملا يستنام إليه.

[9] و قوم جعلوها [الأعداد] مبادئ و لم يجعلوها مفارقة، و هم أصحاب فيثاغورث. و ركبوا كل شيء من الوحدة و الثنائية، و جعلوا الوحدة في حيز الخير و الحصر، و جعلوا الثنائية في حيز الشر و غير الحصر.

> figure. The tradition dates this conception back to Thales and even to the Egyptians. It is certainly very ancient, but it is not seriously attested before the Pythagorean Petron of Himera, who lived in the first half of the fifth century. Knowing the figure thus amounted to knowing not only the figure, but also its properties, as we know in geometry not only the triangle or the square but also the properties that distinguish them. One was satisfied sometimes with ordering the dots into a straight line or sometimes into a square or a rectangle, the dots were not placed on the sides of the figure but on its surface, at the end of the lines hypothetically intersecting in order to form the figure. (Lasserre 1990, p. 57)

Burnet, who confirms Lasserre's analysis,[10] further points out that this way of doing arithmetic was revived by late mathematicians who call themselves Pythagoreans like Nikomachos of Gerasa using the letter alpha instead of points:

> Now it ought to be obvious that this is no innovation, but, like so many things in Neopythagoreanism, a reversion to primitive usage. Of course the employment of the letter alpha to represent the units is derived from the conventional notation; but otherwise we are clearly in presence of something which belongs to the very earliest stage of the science. (Burnet 1908, p. 112)

Ibn Sīnā challenges the Pythagoreans and Neopythagoreans by rejecting the *material* conception of point that underlines their approach which does not distinguish numbers from sensible things, which was the main point of their disagreement with Plato's doctrine:

> According to the ascertained doctrine, however, the point exists only in the line, which is in the surface, which is in the body, which is in matter. The point is not a principle except in the sense of being a limit. In reality, however, it is a body which is the principle in the sense that it is subject to having finitude obtained by it.[11] (p. 254, § 17)

The point of Ibn Sīnā is that points cannot be isolated from lines and lines cannot be isolated from bodies and the latter are in matter, hence points cannot represent the arithmetical unity. Ibn Sīnā's argument is that the existence of a "material point" does not make sense since what actually exists are the bodies; and the points, from which the Pythagoreans claim to compose their figures, should rather be thought of as fictions in which case their doctrine will face the same difficulties as Plato's realism since figures would then be separated from the sensible world.

---

[10] Numbers were represented by dots arranged in symmetrical and easily recognised patterns, of which the marking of dice or dominoes gives us the best idea. And these markings are, in fact, the best proof that this is a genuinely primitive method of indicating numbers; for they are of unknown antiquity, and go back to the time when men could only count by arranging numbers in such patterns, each of which became, as it were, a fresh unit. This way of counting may well be as old as reckoning with the fingers, or even older. (Burnet 1908, p. 111)

[11] و أما على مذهب التحقيق فليست النقطة موجودة إلا في الخط، الذي هو في السطح، الذي هو في الجسم، الذي هو في المادة. و ليست النقطة مبدأ إلا بمعنى الطرف. و أما بالحقيقة فالجسم هو المبدأ، بمعنى أنه معروض له التناهي به.

3.3 Criticism of Aristotle's Conception of Existence in *re* and His Notion of the Infinite

Aristotle's impressive and influential works that included some original contributions to many of the philosophical sciences was, however, far less important in mathematics than the one of his teacher. Mathematics is indeed the discipline where Plato and the Platonists have a strong influence and Aristotle could not afford to be totally silent about it, accordingly he devotes two books, M and N, of his *Metaphysics* to this topic.[12] Julia Annas, who translated the two books in question, points out the briefness of Aristotle's study of the concept of number that is confined to the two books just mentioned:

> Aristotle does not present anywhere in his work a sustained and dialectical attempt to deal with the problems of the philosophy of mathematics (as he does for time, place, etc. in the *Physics*). The positive ideas of his own that are presented in M-N occur in the course of a mainly polemical treatment. (Annas 1976, p. 26)

Let us examine his own views rather than his criticisms of Plato's doctrine which occupy the bulk of the books M and N. Two points are of interest to us: the first is Aristotle's opinion about existence, and the second his alternative solution. The Stagirite sums up his preliminary analysis of existence at the end of Chap. 2, book M:

> It has been adequately shown that mathematical objects are not real objects more than bodies are, that they are not prior to perceptible objects in reality, but only in definition, and that they cannot have separate existence. Since they could not exist in perceptible objects either, clearly either they do not exist at all, or they do so in a certain way and so do not exist without qualification-for we use 'exist' in several senses. (*Metaph.* M. 1077b12-17; in Annas 1976, p. 94)

This is the easy part of Aristotle's analysis; he shows that mathematical objects can exist in no way in actuality for they are not like real objects that can be touched. He still keeps us though in suspense in relation to the question of existence. It is in the next chapter where he reveals what he had in mind:

> The best way of studying each object would be this: to separate and posit what is not separate, as the arithmetician does, and the geometer. A man is one and indivisible as a man, and the arithmetician posits him as one indivisible, then studies what is incidental to a man as indivisible; the geometer, on the other hand studies him neither as man nor as indivisible, but as a solid object. For clearly properties he would have had even if he had not been indivisible can belong to him without them. That is why geometers speak correctly: they talk about existing things and they really do exist-for what exists does so in one of two senses, in actuality or as matter. (*Metaph.* M. 1078a27-31; in Annas 1976, p. 96)

Two surprises can be found in this passage: The first is that he takes objects of geometry as a model of existence. The second is better articulated by Annas:

---

[12] Aristotle has indeed bitterly expressed his dissatisfaction with the Platonists whom he accused of having "turned philosophy into mathematics." (*Metaph.* A 992a32).

> The final phrase gives a new and startling unconnected way of accounting for the way mathematical objects can be truly said to exist even if Platonism is denied. Aristotle here says that numbers, etc. do not *actually* exist, but exist in a different sense. We would expect 'potentially', the usual contrast with 'actually.' At *Physics* 262a8-263b9 Aristotle accounts for the existence of the infinite in this way; it is thus plausible to take the phrase here 'as matter' to mean simply 'potentially'. (Annas 1976, p.151; italics in original)

Crubellier draws the same conclusion: "For Aristotle […] therefore numbers exist only potentially" (Crubellier 1997, p. 98). Perhaps the idea behind is that numbers exist potentially in matter. The precise elucidation of this point is difficult and seems to require further philological and philosophical studies. The main problem is that the pair actuality-potentiality does not help to understand the generation of infinite numbers if numbers are assumed to be potentially *in re*. This is precisely the core of Ibn Sīnā's attack on Aristotle. The attack is twofold: on the one hand he defends the idea that the notion of unity—understood as the result of the fact that one given plurality falls under a concept—is intelligible rather than sensible; and on the other Ibn Sīnā argues that the Aristotelian approach cannot yield the notion of infinite numbers. It is in his *Pointers and Reminders* (*al-Ishārāt wa al-tanbīhāt*), where he clearly and lucidly argues for the compelling (intelligible) existence of the universals and that those universals provide a unity under which falls a given plurality:

> Know that among people's beliefs[13] prevails the following: the existent is the sensible; to assume the existence of what is not accessed (ما لا يناله) by the senses in its substance must

[13]By "people's beliefs (أوهام الناس)", Ibn Sīnā refers to the followers of Aristotle who are not mentioned by name, the explicit identification of the Aristotelians with the common people is made later in *Manṭiq al-Mashriqiyyīn* where he concludes: "as to the common people who engage in such things (أما العامة من مزاولي هذا الشأن), we gave them in the book of *al-Shifā*' what is even too much for them and beyond their requirements" (p. 4). Indeed unlike *al-Shifā'* a work composed for the Aristotelians which explains why there are numerous references to Aristotle as the first teacher whose notion of existence is conspicuously ignored, his attitude seems to have considerably changed in *Pointers and Reminders*. His landmark attack: on Aristotle's notion of existence is so devastating that Goichon has translated the Arabic expression "أوهام الناس" by "vulgaire" (Goichon 1951, p. 351) capturing thus Ibn Sīnā's intended sharp rebuke of the first teacher whose conception of existence turns out to be the belief of ordinary people. This is a clear signal that the Stagirite, who is hardly mentioned throughout, is no longer recognized as an authority. On the contrary, he seems to speak in his capacity as an authority by forcefully emphasizing the "we" in the Arabic sentence "Syllogism, according to what we have established ourselves (على ما حققناه نحن), is…" (Ibn Sīnā 1983, p. 374). In another significant development, which confirms that since *al-Shifā*' Aristotle's views were of little relevance, is his transformation of the notion of proof by adopting *ḥujja* (argument), which was conceived and established as an encompassing epistemic concept of proof by the jurists, to unify the two forms of reasoning: deductive like syllogism (*qiyās*) and non-deductive like induction (*istiqrā'*) and *al-tamthīl* or the *qiyās* of the jurists: "There are three types of argument in establishing something (ثلاثة أصناف ما يحتج به في إثبات شيء…): the first syllogism, the second induction and what accompanies it, and the third *al-tamthīl* and what accompanies it." (ibid., p. 365). As a result, the gap between the author of the *Organon* and the author of *Manṭiq al-Mashriqiyyīn* further widened for the latter remarkably ended up recognising, in *The Second Epistle on the Heavenly Bodies*, the epistemic foundational role of the non-aristotelian *kalām*-type *jadal* (also known as the science of argumentation): "The principles of *all* the sciences are guaranteed by two disciplines (مبادئ العلوم كلها في ضمان صناعتين). From the point of view of proof (أما على السبيل البرهاني), they are guaranteed by the first philosophy which is called divine science; and

therefore be regarded as absurd, and what is not particularised by a location or a position, from itself such as the body, or because of that in which it resides, like the states of the body, has no chance to exist. […] It is possible for you to reflect on the sensible itself and learn from this the falsity of the statements of such people; for both you and he who deserves to be addressed know that these sensible objects fall under (يقع عليها) one and the same name not by pure homonymy, but according to one and the same sense, such as the name human being: neither of you doubts that *Zayd* and *'Amr* fall under it (وقوعه على زيد و عمرو) in the same real sense.[14] […] Therefore the human being, inasmuch as it has a unique reality, or rather inasmuch as its fundamental reality is that in which plurality has no difference, is not sensible but purely intelligible. And the same goes for every universal.[15] (Ibn Sīnā 1968, pp. 7–9)

In fact Aristotle has an alternative formulation based on the notion of measurement:

The measure must always be some identical thing predicable of all the things it measures, e.g. if the things are horses, the measure is 'horse', and if they are men, 'man'. If they are a

(Footnote 13 continued)

from the point of view of conviction (و أما على سبيل الإقناع), they are guaranteed by *al-jadal* [dialectics]. Perhaps the discipline (الصناعة) called in *our time al-kalām* is close to *al-jadal* but less flawed than it." (Ibn Sīnā 1989, p. 41, my emphasis). *Al-Ishārāt*'s powerful impact far exceeded its author's intention since it has triggered one of the important events in the history of thought. For it was commented upon by two major thinkers, the first by the formidable intellectual and *mutakallim* Fakhr al-Dīn al-Rāzī (1149–1209) followed by the famous mathematician and astronomer Nasīr al-Dīn al-Ṭūsī (1201–1274), one of the key members of the Marāgha School which successfully developed the first non-ptolemaic astronomical system in the 14th century. In his commentary of Ibn Sīnā's text, al-Ṭūsī commented upon the comments of his predecessor; both landmark commentaries generated in turn a series of secondary literature on many of the issues discussed by the three great figures.

[14]Ibn Sīnā uses twice the Arabic verb "وقع على" which has the exact meaning of our modern notion "fall under" to capture the relation between individuals and the concept to which they give rise. As this passage already indicates, he establishes instantiation as one of the major concepts of his epistemology and it is not by chance that he mentions just two individuals; we will see why. There is little doubt that this passage, and others as we shall see, bears the influence of one of al-Fārābī's important mathematical works *Kitāb al-wāḥid wa al-waḥda* or *The Book on the One and the Unity* which turns out to be crucial to the development of our author's conception of number. The philosopher of Baghdad begins indeed by explaining in what way the one can be said many as follows: "we say in many circumstances that two things are the same, and that this and that are the same there is no difference between them." (al-Fārābī 1989, p. 36 §1) Then he immediately discusses examples in which two can be said one and the same.

We only say that either the nearest genus of each two things is one and the same like our saying the donkey and the horse are one in animality if animal, which is their genus, is one and the same; or either the last species of each two things is one and the same like our saying *Zayd* and *'Amr* are one in humanity if human–their nearest species–is one and the same. (ibid., p. 37 § 2)

[15]إعلم أنه قد يغلب على أوهام الناس أن الموجود هو المحسوس، و أن ما لا يناله الحس بجوهره، ففرض وجوده محال، و أن ما لا يتخصص بمكان أو وضع بذاته كالجسم، أو بسبب ما هو فيه كأحوال الجسم، فلا حظ له من الموجود. و أنت يتأتى لك أن تتأمل نفس المحسوس، فتعلم منه بطلان قول هؤلاء؛ لأنك، و من يستحق أن يخاطب، تعلمان أن هذه المحسوسات، قد يقع عليها اسم واحد، لا على سبيل الإشتراك الصرف، بل بحسب معنى واحد، مثل اسم الإنسان: فإنكما لا تشكان في أن وقوعه على زيد و عمرو بمعنى واحد موجود [...]. فإذن الإنسان، من حيث هو واحد الحقيقة، بل من حيث حقيقته الأصلية التي لا تختلف فيها الكثرة، غير محسوس بل معقول صرف. و كذلك الحال في كل كلي.

> man, a horse, and a god, the measure is perhaps 'living being', and the number of them will be a number of living beings. If the things are 'man' and 'pale' and 'walking', these will scarcely have a number, because all belong to a subject which is one and the same in number, yet the number of these will be a number of 'kinds' or of some such term. (*Metaph.* N. 1088a7–1088a14; in Annas 1976, p. 117)

Thus, instead of understanding unity as result of falling under a concept as expressed by Ibn Sīnā, the author of the *Organon* suggests the linking of the notion of number with the unit of measurement, called "numerical number" (Crubellier 1997, p. 99) or as Annas puts it "number is relative to what is numbered" (Annas 1976, p. 39). Among several criticisms of the notion of numerical number, Ibn Sīnā launches the following attack:

> It follows necessarily for those upholding the doctrine of numerical number, composing from it the forms of natural things, to do one of the two things: they must either make for the separated existing number a finite termination—in which case its finitude would obtain one limit rather than any other of the invented limits that remain unrealized—or make it infinite, thereby rendering the forms of natural things infinite.[16] (pp. 251–252, § 9)

In other words, making numbers dependent on measuring empirical objects fails to capture their real nature: the infinite.[17] It seems thus that Aristotle denies our ability to indefinitely count numbers independently of empirical objects.

Ibn Sīnā's systematic survey of the doctrines of the ancients shows that the philosophy of mathematics, like natural philosophy for that matter, had been in crisis not to say in disarray since the 4th century B.C., and it seems that the three schools, the Pythagoreans, the Platonists and the Aristotelians, were little aware of it. As a result, neither Greek philosophers nor mathematicians were able to break the long period of scientific stagnation despite the rivalry between the three antagonists. For things to change, we have to wait until the 9th century, the distinguished historian of mathematics Roshdi Rashed reminds us of the memorable scientific event that reinvented mathematics in Bagdad, the famous capital of the ruler Hārūn al-Rachīd:

> Al-Khwārizmī's book of Algebra is one of those works which has shaped the fate of mathematics. Written in the beginning of the 9th century in Bagdad, translated three times in Latin since the 12th century then a little a bit later in Italian, its influence was

---

[16] و يلزم القائلين بالعدد العددي المركبين منها صور الطبيعيات أن يعملوا أحد شيئين: إما أن يجعلوا للعدد المفارق الموجود نهاية، فيكون تناهيه عند حد من الحدود دون غيره من اختراع الذي لا محصول له، أو يجعلوه غير متناه فيجعلوا صور الطبيعيات غير متناهية.

[17] We have seen how by al-Fārābī's time arithmetic has already been conceived as a pure science, in a significant passage Ibn Sīnā further mentions its a priori epistemic status: "Someone, however, may say: the things of pure mathematics (الرياضية المحضة) which are examined in arithmetic and geometry, are also prior to experience—particularly number, for there is no dependency at all for its existence on experience (و خصوصا العدد فإنه لا تعلق لوجوده بالطبيعة ألبتة) because it cannot be found in experience." (p. 17, § 14). This specificity of arithmetic is stressed by many 19th century mathematicians like Gauss who strikingly expressed a similar view in his letter to Olbers (1817) following the discovery of non-Euclidean geometries: "geometry must not stand with arithmetic which is purely a priori" (Gauss 1900, vol. VIII, p. 177). Ibn Sīnā would wholly agree with Gauss since for him the concept of number is so pure that even time is not essential to its construction.

> continuously felt across the universal mathematical thought. Al-Khwarizmi's book is foundational in many respects: first, in algebra, as an independent mathematical discipline; then in many applications of algebra afterwards, in arithmetic and geometry, which continue to enrich mathematics with many new chapters, to deeply reshape their configuration in order to finally impose a new rationality: algebraic and analytic. (Rashed in Al-Khwārizmī 2007, from the back cover)

In the same vein, Lasserre sums up the quiet mathematical revolution that took 1300 years to see the light of day:

> While modern arithmetic mainly developed the art of computing and explored by the means of algebra the theory of relations between numbers, ancient arithmetic's task was to 'know the numbers'. (Laserre 1990, p. 56)

According to our view, our physician has fully grasped the major formal structural shift brought about by the founder of algebra from theory of objects to theory of relations, a mathematical revolution that induced in turn an epistemological revolution; where the epistemological point is the elucidation of epistemological concepts that should help the development of mathematics itself including pure and applied mathematics.

3.4 Criticism of Euclid's Conception of Plurality

Crubellier argues that the definition of number in the *Elements* has a fascinating history which ends up with Euclid being sided with Aristotle.

> The treatise on arithmetic which occupied the books VII, VIII and IX of the Euclidean Elements begins with the definitions of unity and number. The latter is defined [VII, def. 2] as 'a multiplicity composed of units'; and unity is "that by virtue of which each of the things is called one' [def. 1]. We can legitimately be puzzled by this definition. [...] It is because it has a history. The definition condenses in a few words the results of discussions and debates that took place in the centuries preceding Euclid on the nature of numbers and their significance for understanding and explaining the world. One of the important moments of this tradition of the philosophy of number has been, in the 4th century BC, the Platonic doctrine of numbers-ideas and its criticism by Aristotle; and we shall see that the way Euclid defines unity amounts to siding with Aristotle's point of view. (Crubellier 1997, p. 81)

It looks as if Aristotle, despite the many difficulties of his approach outlined above, has triumphed over Plato and his followers.[18] And yet Aristotle's definition does

[18]In her general comment on Chap. 2 of book M, which deals with existence, Annas agrees when she declares that "Aristotle demolishes the Platonist conception of mathematical objects before offering his own alternative in Chap. 3" (Annas 1976, p. 137). If that is the case, i.e. if Aristotle's scattered views on the philosophy of mathematics were successful in overthrowing the powerful realist metaphysical doctrine of his master, it is only ironic to find out that Ibn Sīnā, who is regarded by most scholars as one of the great Aristotelians, has actually performed the first break with Aristotle's conceptual apparatus in a fundamental chapter of theoretical philosophy; a break that took 1500 years to come to fruition!

not seem to be working. Our author contests the definition in a way that strongly reminds us of Frege's criticism of the notion of number as plurality:

> One is astounded by those who define number and say, "number is a plurality composed of units or of ones," when plurality is the same as number (و الكثرة نفس العدد) and the reality of plurality consists in that it is composed of units. Hence, their statement, "plurality is composed of units," is like their saying, "plurality is plurality (إن الكثرة كثرة)." For plurality is nothing but a *name* for that which is composed of units (فإن الكثرة ليست إلا اسما للمؤلف من الوحدات).[19] (p. 80, § 5; my emphasis)

In Book 3 Chap. 5, Ibn Sīnā states what Euclid might perhaps mean by his definition: "the definition of each number—if you wish true ascertainment—is to say:

> It is number formed from the combination of one and one and one,' mentioning all the ones (إنه عدد من اجتماع واحد و واحد و واحد، و تذكر الآحاد كلها)." (p. 93, § 6)

Certainly, this explanation assumes that "one" is given. It is misleading thus to believe that the Euclidean definition defines the general concept of number even if it is assumed that "one" does not refer to an empirical entity,[20] what it can at best define is only particular numbers by mentioning all the ones of each number. A cumbersome conception which was finally overcome by the invention of the decimal positional system to represent the process of generation of all numbers:

[19]By stressing that a number is just a name, Ibn Sīnā clearly excludes that it could be the property of things like colour as Frege argues in this passage: "It marks, therefore, an important difference between colour and Number, that a colour such as blue belongs to a surface independently of any choice of ours. [...] The Number 1, on the other hand, or 100 or any other Number, cannot be said to belong to the pile of playing cards in its own right, but at most to belong to it in view of the way in which we have chosen to regard it; and even then not in such a way that we can simply assign the Number to it as a predicate" (Frege 1884, p. 29, § 22). And he further says: "Some writers define Number as a set or multitude or plurality. [...] These terms are utterly vague: sometimes they approximate its meaning to "heap" or "group" or "agglomeration", referring to a juxtaposition in space, sometimes they are so used as to be practically equivalent to "Number", only vaguer. No analysis of the concept of Number, therefore, is to be found in a definition of this kind." (Frege 1884, p. 38, § 28).

[20]Frege, who quotes Euclid's definitions verbatim i.e. in Greek, indeed points out the ambiguity of the definition of unity: "In the definitions which Euclid gives at the beginning of Book VII of the Elements, he seems to mean by the word '*μονάς*' sometimes an object to be numbered, sometimes a property of such an object, and sometimes the number one" (Frege 1884, p. 39, § 29); a confusion that will be confirmed by Ibn Sīnā, as we shall see, by criticising those who try to explain the generation of numbers through repetition by conceiving unity as a metaphysical property rather than a principle of the composition of number.

> To consider number in terms of its ones, however, is among the difficult things for both the imagination and verbal expression, so that one resorts to descriptions.[21] (ibid.)

[21]This is not the only passage where Ibn Sīnā establishes a relevant epistemic connection between objects like the natural numbers and the means used to represent them. Let us recall that he extensively studied Euclid's *Elements* and he wrote an arithmetic book he called *al-Arithmāṭīqī* with the aim of further advancing the state of the art, an unprecedented rich mathematical experience in the formation of a philosopher that enabled him to raise important epistemic and pragmatic questions as result of his awareness that the surge of mathematics since the ninth century was due to the emergence of a new scientific approach which bound theory to action through the notion of benefit and usefulness. A major shift in the conception of theoretical research whose aim is not so much to produce a rival philosophical discourse to defeat one's opponent but rather to come up with a useful theory that could help to advance scientific practice. Ibn Sīnā's reflection is particularly relevant to the Greek approach to mathematics. What is for example the relation between Euclid's definitions of the concept of number, by which according to the classicists Aristotle defeated the philosophy of his master, and his mathematical practice? What kind of representation did he actually use to perform his calculations? And to what extent did his definitions reflect or impact his mathematical practice? If numbers are for example conceived as combination of ones, how can they thus be represented? This was one of the hardest problems faced by Greek mathematicians, how to represent for example the cumbersome conception "one and one and one"? The various historical answers to these and similar basic epistemic questions had major implications for the development of mathematics and with it for the rest of science. For the symbolic representation of numbers is part of the intelligibility of their conception and the conception of their relation and manipulation as pointed out by van der Waerden in his Chap. 2 where he presents "a brief survey of the number system and the number notations in the principal cultural periods, and of the related arithmetical techniques," he points out:

> We shall see that these notations and these techniques are of great importance for the development of mathematics; not, of course, in the sense that a good number system leads automatically to a high development in mathematics, but rather that a good notation and a convenient manipulation of the four fundamental operations are necessary conditions for the development of mathematics. Without mastery of these fundamental operations, mathematics cannot get beyond a *certain low level*. This shows itself most clearly in algebra. (van der Waerden 1961, p. 37; my emphasis)

Little helped by the doctrine of the philosophers who preferred to speculate more on natural philosophy, as pointed out by Ibn Sīnā, than tackle basic concrete theoretical problems in mathematics such as the symbolic representation of numbers, Hellenic and Hellenistic mathematicians found it hard, throughout their long supreme reign over the mathematical realm, to conceive a symbolic system to represent the natural numbers which was one of the main stumbling blocks to the development of number theory and algebra. They particularly failed to reflect and improve upon the positional sexagesimal system of the Babylonians that has marvellously worked in measuring time we still use today and by which astronomy was able to take off and make such remarkable progress. They adopted instead a backward alphabetical notation which was inadequate to capture the concept of number as structure as the great Dutch historian of mathematics explains:

> The Greek notation for numbers as compared with the excellent Babylonian notation, was really a retrogression. In most remote antiquity, they had a notation, which resembles the well-known Roman numerals. The letters *Π, Δ, H, X, M* are of course the initial letters of the Greek words for 5, 10, 100, 1000 and 10,000. Later on, a briefer, alphabetical notation was introduced. To distinguish numbers from words, an accent was added at the end, or a dash was placed over them, such as

Furthermore, Ibn Sīnā thinks that any definition of number is doomed to circularity for it presupposes the apprehension of the concepts by which number is defined[22]—the only way out as discussed in the following chapters is to introduce a specific mental act that accounts for the construction of numbers:

> Those who believe that they escape from the difficulty of circular definition once they say, "Number is a discrete quantity that has order," do not escape it. For the soul's conception of quantity requires that quantity be known in terms of part, division, or equality. As for part and division, these can be conceived only through quantity. As for equality, quantity is better known to the unclouded intellect because equality is one of the accidents proper to quantity, an accident in whose definition quantity must be included. To this it is said: equality is unity in quantity and order, which are also in the definition of number. It is one of the things that is understood only after number is understood. (p. 81, § 8)

(Footnote 21 continued)

$$\overline{\alpha\tau\varepsilon} \text{ or } \alpha\tau\varepsilon' = 1305$$

> Numbers beyond the myriad $M = 10^4$ were designated by the use of the symbol M, e.g.:
>
> $$\overset{\kappa\varepsilon}{M}\,\mu\gamma' = 250043$$
>
> In place of $\overset{\kappa\varepsilon}{M}$ one could also write $\ddot{\kappa}\ddot{\varepsilon}$ For higher powers of M, Archimedes and Apollonius used still different notations. The use of letters for specified numbers was not advantageous for the development of algebra. Until the time of Plato's friend Archytas (390 B.C.), letters were used for indeterminates; in Archytas, $\Gamma\Delta$ represented for instance the sum of the numbers Γ and Δ. If supplemented by a sign for a multiplication, a minus sign and a symbol for fractions, this system might have provided an effective notation for theoretical arithmetic. But even Euclid (300 B.C.) had already abandoned this simple notation for sums, probably I think, to avoid confusion with the alphabetical number symbols. When Euclid wants to add two numbers, he represents them by means of line segments *AB* and *BΓ*, and denotes the sum by *AΓ*. For purposes of calculation, the Greek number symbols were about equally troublesome. (van der Waerden 1961, pp. 45–46)

[22]Poincaré, who rejects similar attempts to define numbers at the end of the 19th century and early 20th century, better articulates the circularity argument that Ibn Sīnā has in mind:

> Les définitions du nombre sont très nombreuses et très diverses; je renonce à énumérer même les noms de leurs auteurs. Nous ne devons pas nous étonner qu'il y en ait tant. Si l'une d'elles était satisfaisante, on n'en donnerait plus de nouvelle. Si chaque nouveau philosophe qui s'est occupé de cette question a cru devoir en inventer une autre, c'est qu'il n'était pas satisfait de celles de ses devanciers, et s'il n'en était pas satisfait, c'est qu'il croyait y voir une pétition de principe. J'ai toujours éprouvé, en lisant les écrits consacrés à ce problème, un profond sentiment de malaise ; je m'attendais toujours à me heurter à une pétition de principe et, quand je ne l'apercevais pas tout de suite, j'avais la crainte d'avoir mal regardé. C'est qu'il est impossible de donner une définition sans énoncer une phrase, et difficile d'énoncer une phrase sans y mettre un nom de nombre, ou au moins le mot plusieurs, ou au moins un mot au pluriel. Et alors la pente est glissante et à chaque instant on risque de tomber dans une pétition de principe. (Poincaré 1905, p. 821)

Ibn Sīnā criticises the Euclidean definition of number not only because it is circular, but more importantly it can be seen as inadequate for it conceals in fact the real function of numbers in mathematical practice. For if number is reduced to plurality, how can a plurality then be a mathematical object?

> For each of the numbers there is a reality proper to it and a form in terms of which it is conceived in the soul. This reality is its unity, by virtue of which it is what it is. Number is not a plurality that does not combine to form one unity, so as to say, 'it is simply as an aggregate of ones.' For, inasmuch as it is an aggregate, it is a unit bearing properties that don't belong to another. It is not strange for a thing to be one inasmuch as it has some form (for example, being ten and being three) and yet possess plurality. For, with respect to being ten, it is what it is by virtue of the properties belonging to ten. As for its plurality, it possesses only the properties belonging to plurality in opposition to unity. For this reason, then, ten, in being ten, does not divide into tens, each of them possessing the properties of being ten. (pp. 91–92, § 4)

It is this unity aspect that explains the role of numbers as mathematical objects, and enables the working mathematician to attribute some properties to them, prove his statements and perform his calculations.

Ibn Sīnā further points out that it is precisely the conception of numbers as mathematical objects that explains the inadequacy of the grammatical conjunction to render the meaning of elementary arithmetic equations:

> It must not be said that ten is nothing but nine and one, or five and five, or that is one and one and so on until it terminates with ten. For your statement, "ten is nine and one", is a statement in which you predicated nine of the ten, conjoining the one with it. It would be as though you had said, "ten is black and sweet", where, by the descriptions, the one conjoined to the other must be true of the statement; thus the ten would be nine and also one. If, by the conjunction, you did not intend a definition but instead intend something parallel to the statement, "man is an animal and is rational"—that is, "man is an animal: that animal which is rational"—it would be though as you had said, "ten, is nine, that nine which is one", and this is also impossible. (p. 92, § 5).

The point of our author is that the grammatical conjunction, whose function is rather to list, does not explain the composition of a number. In fact, Ibn Sīnā lucidly understands an operation like sum as a *transformation* of one side of the equation into the other[23]:

---

[23]It is this grasping of elementary arithmetic processes like sum as operators which will be crucial to his account for repetition by remarkably making it the most basic mathematical operator. It is important to point out that Ibn Sīnā does not conduct this linguistic analysis by chance for he explicitly mentions that the analysis of language is among the tasks of the logician: "because there is some link between the word and the meaning, the modes of the word can influence the modes of meaning. That's why the logician should also consider the absolute [semantic] aspect of the word inasmuch as it is not restricted by the language of a certain people to the exclusion of other, except in rare cases" (Ibn Sīnā 1983, p. 131). In *Manṭiq al-Mashriqiyyīn*, he becomes closer to the pragmatist position of the linguists for he begins with this topic in which logical analysis is tightly linked to the discussion of the meaning of sentences. A prominent example is his analysis of Arabic sentences, he explains their structure by admitting the existence of a second type of propositions which are made of just two components and he points out that the copula is not needed to link the subject to the predicate:

> Rather, ten is the sum of nine and one when taken together, they are transformed (صار) to something other than either.
>
> بل العشرة مجموع التسعة و الواحد إذا أخذا جميعا فصار منهما شيء غيرهما. (p. 92, § 5)

Ibn Sīnā's general refutation of the concept of number in the Greek tradition shows that he is aware of introducing a major shift in the philosophy of mathematics which is neither Platonist nor Aristotelian, providing the necessary evidence that vindicates his claim of breaking with the past first announced in his introduction of *al-Shifā'* and then effectively enforced in his last work *Manṭiq al-Mashriqiyyīn.* In the rest of the book, we will discuss his own conception of numbers. From al-Fārābī he learns that numbers are objects of the mind and one of the difficult questions in this regard is how can a number such as 7 e.g. be an object of the mind? In his attempt to answer questions like these, Ibn Sīnā has developed his theory of knowledge in which numbers are conceived as intentional objects. His intentionality is widely conceived so as to include irrational numbers among the objects generated and accessed by the mind; a mathematical conception that seems to have physical implications by making mathematical structures one of the main means by which the mind has some grip on the outside world.

(Footnote 23 continued)

> إذا كانت القضية غير ثلاثية، إنما هي ثنائية فقط لم تذكر فيها الرابطة استغناء، لأن محمولها كلمة أو اسم مشتق اشتقاقا يتضمن النسبة المذكورة على حسب اللغة (Ibn Sīnā 1910, p. 67).

Arabic sentences are no longer regarded as odd propositions and consequently logically ill formed formulas due to the absence of the copula as it seems to be considered by the logicians during al-Fārābī times; it becomes an exception (a rare case) that can provide an insight into the formalisation process of sentences for our author no longer requires that the intentional relation between the subject and the predicate in Arabic sentences be formally expressed. He is thus able to discuss the meaning of propositions from the perspective of the Arabic language without introducing the copula since he formalised it simply as "b c (ب ج)" where b designates the subject and c designates the predicate (ibid. p. 64, also p. 68 and pp. 61–62).

Chapter 4
Ibn Sīnā's Basic Theory of Knowledge

4.1 Intentionality as Mental Existence

One decisive point in Ibn Sīnā's strategy is to acknowledge that numbers are objects. Furthermore, if they are objects they possess (some kind of) reality since

> If you were to say: 'The reality of such thing is a thing' this would not be a statement imparting knowledge of what is not known. Even less useful than this is for you to say: 'Reality is a thing' unless by thing you mean "the existent"; for then it is as though you have said "The reality of such a thing is an existing reality. (p. 24, §11)
>
> The meaning of existence is permanently concomitant with it because the thing exists either in the concrete or in the estimative faculty and the intellect, if this were not the case, it would not be a thing. (p. 25, §11)

Now the question is what kind of reality or existence is involved in the notion of number?

> Number has an existence in things and an existence in the soul. The statement of one who says that number exists only in the soul is not reliable. But should he say that number stripped from what is counted (مجردا من المعدودات) that are in concrete existence has no existence except in the soul, this would be true. (p. 91, §2)

This paragraph must be read carefully and linked with the next parts of Ibn Sīnā's conception of number. Numbers are not only an abstraction from some properties but they have a separate existence in the mind. As we will see, they are intentional objects, the products of a specific intentional act that makes it possible to generate objects beyond the sensible experience such as infinite numbers.

Thus the necessary condition for something, which does not exist in the external world, to be an object is to exist in the mind. Ibn Sīnā draws our attention to the

H. Tahiri, *Mathematics and the Mind*,
SpringerBriefs in Philosophy, DOI 10.1007/978-3-319-25238-4_4

specificity of our mind: its content is an object whose presence is a specific form of existence. It is in the nature of our mind to have some content as object for to think is to think about something, what Ibn Sīnā calls then existence in the mind or mental existence (الوجود الذهني) is the simple presence of an object in the mind, "this is the meaning intended by the thing" he concludes (p. 25, §11). This is a major ontological extension that provides a general solution to the mystery of how we can meaningfully talk about a very large class of strange objects that have no real existence like fictions. Ibn Sīnā seems to have discovered that the real relation between the mind and its content is simply intentional: the objects of our intentional acts need not be physical, spatiotemporal, or ideal entities, and they need not exist independently of our intentional acts. An idea that he clarifies further by rejecting the claim defended by some faction of *al-mutakallimūn* (theologian-philosophers) that non-existent is a thing:

> If by the nonexistent is meant the nonexistent in external reality, this would be possible; for it is possible for a thing that does not exist in external things to exist in the mind. But if something other than this is meant, this would be false and there would be no information about it at all. It would not be known except only as something conceived in the soul. To the notion that the nonexistent would be conceived in the soul as a concept that refers to some external things, we say certainly not.[1] (p. 25, §12)

[1]This text appears to make irrational numbers nonintentional objects which assumes that they are part of the nonexistents, which is not the case for the following reason: "reference to the nonexistent that has no concept in any respect at all in the mind is impossible" (p. 25, §13). Ibn Sīnā tells us here why nonexistents are not intentional objects: the words used seem to refer to something while in fact they do not refer at all that is why it is impossible for the mind to use means by which to have access to what they refer; an example of such nonexistents is the squared circle or the set of all sets that are not members of themselves. To distinguish irrational numbers and nonexistents, Ibn Sīnā uses one of his key epistemic concepts, the notion of accessibility relation, in this significant passage: "there are, here, relationships in irrational roots and in numerical relations that are easily accessible (قريبة المنال) to the soul." Unlike nonexistents, irrational numbers, though they fail to refer to external things, are within reach of the mind for it can actually "see" the limit of their infinite development. As a result of the apprehension of their convergence, irrational numbers can be accessed and manipulated by the mind through symbols and numerical relations. *Al-Manāl* or accessibility, in the sense of within reach, is thus a specific mental act by which the mind can grasp and refer to a determined ongoing open process that does not need to be (assumed) actually realised. It is not though clear what Ibn Sīnā astonishingly means by "there are here." The next sentence appears to say that what he has in mind is the manipulation of irrational numbers due to the successful extension of arithmetical operations to them by al-Khwārizmī. If it is the case, the statement can then be seen as the conclusion drawn by Ibn Sīnā in which he seems to adopt some kind of structural epistemic realism in physics according to which the human mind has access to the external world through mathematical structures by constructing relations that are satisfied by some class of objects, as suggested by the following passage: "knowledge (معرفة) of the order of the arrangement of the spheres can only be arrived at through astronomy (علم الهيئة); and astronomy is only arrived at through the science of arithmetic and geometry." (p. 15, §6).

This significant passage clarifies what seems to be the mysterious relation established earlier between "existence of number in things" and "existence of number in the soul." For Ibn Sīnā, numbers should be conceived as intentional objects for they refer to some extramental entities.[2] Numbers like 10 e.g. exist in our mind as intentional objects as a result of being stripped from what is counted but also exist in the outside world for they refer to particular objects since we can at any time present things that serve as referents, like fingers for example, whose total is ten. Ibn Sīnā's intentionality seems to play then a surprisingly double function: the first is mental existence by stripping numbers from what is counted, and the second is

[2]The theory of intentionality in the Arabic tradition is another chapter of research yet to be investigated. The jurists are behind its development since questions about the meaning of legal rules and their application to individual cases are an essential part of the discussion between the parties in *al-jadal* (Miller 1985, Chap. I, p. 15). For example, in his discussion of one of the specific technical attacks, al-Juwainī (1028–1085) points out that for the jurists "the meaning is what is intended by i.e. the reference of the verbal sentence (المعنى هو المقصود باللفظ)" (al-Juwainī 1979, p. 212). This is the usual meaning of *ma'nā* defined by the lexicographers such as *Ibn Manẓour* in his *Lisān al-'arab*: "وَمَعْنَى كلّ كلامٍ و معناته ومَعْنِيَّته: مَقْصِدُه، و الإِسْم العَناء. وعَنَيْتُ بالقَول كَذا، أي أردتُ و قصدتُ." i.e. "the *ma'nā* of any discourse is its intended meaning, and *'anaytu* by the saying this' is 'what I meant and intended'". Abū 'Abd Allah ibn Ziād better known as Ibn al-A'rābī (767–845), one of the early leading linguists, is more precise when he defines *ma'nā* as: "the intention (القصد) which comes out and appears in the thing when it was searched for" (in Ibn Fāris 1984, Vol. IV, pp. 148–149). It is al-Juwainī's disciple al-Ghazālī who fixed the meaning of intentionality as *qasd* and showed its basic epistemic function in any discourse through his popular books like *Maqāsid al-falāsifa* or *The Intentions of the Philosophers*. He also establishes its use in law and jurisprudence making it a general methodological interpretation of legal texts famously known as *maqāsid al-sharī'a*. As for Ibn Sīnā, he not only identifies the mind with intentionality by declaring: "every intention tends towards something (إن كل قصد فله مقصود)" (Ibn Sīnā, book 9, Chap. 3 §7, p. 320), repeated almost exactly in another passage: "every intention is designed for the intended object (كل قصد يكون من أجل المقصود)" (ibid. §5, p. 319). He further specifies that meaning is not transparent to the mind, it is rather determined by the will of the enunciator to convey his intention: "the verbal word by itself has no meaning… it only acquires its meaning by the will of the enunciator (اللفظ بنفسه لا يدل ألبتة إنما يدل بإرادة اللافظ)" (Ibn Sīnā 1952, p. 25). Though intentionality is not Ibn Sīnā's innovation, it is how he integrates it in his epistemology by using it in correlation with abstraction that should be appreciated. For it is particularly remarkable how he swiftly gets to intentionality through his conception of abstraction. As it will be explained in the next section, his brilliant idea is to conceive of abstraction as that mental operation which makes the mind an object of study for it is the locus of complex mental activities whose analysis can lead to the distinction of those intentional acts involved in the construction of the concept of number. The scholastics translation of Arabic words (*ma'nā*, *murād*, *qaṣd*) as *intentio* seems to only stress the presence in the mind aspect as suggested by Husserl's master Brentano in his *Psychology from an Empirical Standpoint*'s famous passage:

> Every mental phenomenon is characterized by what the Scholastics of the Middle Ages called the intentional (or mental) inexistence [i.e. existence *in intentio*] of an object, and what we might call, though not wholly unambiguously, reference to a content, direction toward an object (which is not to be understood here as meaning a thing), or immanent objectivity. Every mental phenomenon includes something as object within itself, although they do not all do so in the same way." (Brentano 1995, p. 88)

what seems to be the reverse function by referring (back) to particular objects in the outside world. However, as already mentioned, the back reference is not always a simple inverse one to one relation, particularly not in relation to the generation of infinite numbers as this can be launched by a sensible experience which triggers a repetition process that goes beyond that sensible experience. This double mental operation, which explains the powerful creativity of the human mind, raises nevertheless more questions: for how can numbers be stripped from what is counted to become intentional objects? How many numbers can be formed in this way i.e. by abstraction? And once abstracted, how can that which is counted and separated form nevertheless a single entity in the mind? And how can such an abstracted entity refer back to particular objects of the outside world? To tackle these complex issues, Ibn Sīnā simply reinvents epistemology in which intentionality acts as an interaction between the mind and the world and then distinguishes an act of intentionality specific to the generation of numbers.

Ibn Sīnā outlines his basic intuitive theory of knowledge in *Pointers and Reminders*[3]:

> Sometimes an object is sensible (محسوسا), and that is when it is seen; then it is re-presented (متخيلا), when it is absent, by internally presenting its form; like when you see *Zayd* for example and you re-present him when he is absent. Sometimes an object is intelligible (معقولا), and that is like when you understand the meaning of human from *Zayd*, a meaning that holds for other things as well. When *Zayd* is sensible, he is covered with veils, foreign to his quiddity, which do not affect his real quiddity if they were removed, like to have place, position, quality and determined quantity, such that *if he was substituted by imagination (تُوُهِّمَ) with someone other than him, the reality of human's quiddity would not have been affected.*[4] Sense (الحس) will get access (يناله) to *Zayd* inasmuch as he is enveloped in these accidents attached to him because of the matter from which he is created. Sense does

[3]This section can be seen as an indication of the kind of philosophy that Ibn Sīnā describes "as it is in nature" which can best be illuminated by "intuitive" as used in the computer software sense, i.e. easy to understand and use. Hence by intuitive theory of knowledge we mean a concrete descriptive account as experienced by the epistemic subject.

[4]The substitution of elements of an equivalence class is clearly expressed by al-Fārābī in the following passage:

> And we can also say in each of the two things in virtue of each one of them leads to the same purpose that they are all the same. We therefore say regarding their plurality *use whatever you wish since they are both one and the same* استعمل أيهما شئت فكلهما واحد. (Al-Fārābī 1989, p. 38 §3; my emphasis)

He also examines the case of substitution of names that refer to the same object, he calls this identity relation "*wāhid bi al-'adad* i.e. one in number" or more significantly "*wāhid bi 'aynihi* i.e. one in itself." One of the interesting examples he provides is the customary use of name and *kunya* to refer to the same person (ibid., p. 41 §6); Ibn Sīnā provides the following specific example in his *al-Ilāhiyāt*: "*Zayd* and *Ibn 'Abdallah* (i.e. *'Abdallah*'s son) are one" (Ibn Sīnā 2005, p. 74 §2), we will come back to this in the last section of this chapter.

> not remove them from *Zayd* and will gain access to him only (لا يناله) in a positional relation between his perception and his matter; that's why his form is not re-presented in the external sense (الحس الظاهر) when he is away.[5] (Ibn Sīnā 1992, pp. 367–369)

Three regions are mentioned here by Ibn Sīnā: (1) sensibility (*al-iḥsās*), (2) memory (*al-khayāl*), (3) the intellect (*al-'aql*).[6]

4.2 Sensibility (*al-iḥsās*)

The above passage makes clear why experience is the foundation of knowledge. Hence the importance of sensibility since it is through our senses that we have access to the outside world. We cannot know what is a colour like green if we have never seen it or a piece of melody such as *Rubā'iyyāt al-Khayyām* if we have never heard it and even by knowing them we cannot talk about them to people who don't share the same experiences. This region of the mind corresponds thus to our daily life in which we are dealing essentially with concrete objects. We pay more attention (تأمل) to the individual, and the mind tends to analyse more to avoid confusing two individuals. As Ibn Sīnā puts it, "Analysis is to distinguish things whose existence truly is in the composite but they appear confused to the mind (إن التحليل تمييز أشياء صح وجودها في المُجْتَمَع و لكنها مختلطة عند العقل)"[7] (Avicenna 2009, Book 2 Chap. 9, p. 209). Mistakes are inevitable due to the overwhelming amount of concrete data to be processed.

4.3 Memory (*al-khayāl*)

Ibn Sīnā defines *al-khayāl* as the faculty that memorises (*taḥfiẓu*) re-presentatives of all what is accessed by the senses (*muthala al-maḥsusāt*) after its absence, gathering in it (*mujtama'a fīhā*) memories of all things, state of affairs and events previously

---

[5] الشيء قد يكون محسوسا، عند ما يشاهد، ثم يكون متخيلا عند غيبته ، بتمثل صورته فى الباطن، كزيد الذي أبصرته ، مثلا، إذا غاب عنك فتخيلته. قد يكون معقولا عند ما يتصور من زيد ، مثلا، معنى الإنسان الموجود أيضا لغيره. وهوعندما يكون محسوسا يكون قد غشيته غواش غريبة عن ماهيته ، لو أزيلت عنه لم تُؤَثِّر في كنه ماهيته، مثل : أين، و وضع، و كيف، و مقدار بعينه؛ لو تُوُهِّمَ بَدَلَه غَيْرُهُ لم تُؤَثِّر في حقيقة ماهية إنسانيته. والحس يناله من حيث هو مغمور في هذه العوارض التي تلحقه بسبب المادة، التي خلق منها، لا يجرده عنها ولا يناله إلا بعلاقة وضعية بين حسه ومادته، ولذلك لا يتمثل في الحس الظاهر صورته إذا زال.

[6] The term "regions" seems most appropriate as "stages" or "levels" are inappropriate for they connote some kind of ladder-like hierarchical structure of the mind making difficult to speak of their interaction. This way of presenting this distinction is to signal three kinds of object correspond to the three main regions of the mind: memory, imagination and the intellect.

[7] التحليل تمييز أشياء صح وجودها في المجتمع و لكنها مختلطة عند العقل، فيفصل بعضها عن بعض بقوته و بحده. أو يكون بعضها يدل على وجود الآخر، فإذا تأمل حال بعضها انتقل منه إلى الآخر.

experienced.[8] *Al-kahyāl* is then the memory of sensibility for when the concrete object is no longer present to the senses, memory comes to our rescue by re-presenting it to our mind. The power of memory is very limited, however, for it can never re-present the entire concrete object; only a small part of it can be recalled. It is in this sense that memory represents the first and the most important level of *tajrīd* or abstraction since what remains is actually just a tiny fraction or a sketch of the concrete object.

Al-khayāl is not thus imagination, the usual word used to render *khayāl*, since what is experienced cannot be imagined but rather re-presented. *Al-khayāl* is thus that essential region of the mind without which imagination cannot work since it re-presents to the mind the stored sketches (الخيالات) in the absence of the corresponding sensible objects; he also calls it *al-muṣawwira* (المصوِّرة) in the sense of *al-ḥāfiẓa li-ṣṣuwwar* (الحافظة للصور) as he specifies[9] i.e. the memory of sketches imprinted by the internal senses. Though the stored sketches are not the whole memory, they nevertheless represent its most basic part identified by Ibn Sīnā as *al-khayāl* due precisely to the foundational role of its specific content in the construction of knowledge, for it is entirely made of sketches of individuals i.e. particular space-time located objects that were actually accessed and perceived by the senses. For short we call *khayāl* memory instead of memory of sensibility since there is no risk of confusion as it is the only faculty needed in the construction of numbers.

As for imagination which is called by Ibn Sīnā (المتخيلة) due to its active nature, it is a second degree faculty for he describes it as the faculty which is capable of acting on objects of memory by performing various operations by combining and separating abstracted entities in a variety of ways. The interesting feature of Ibn Sīnā's epistemology is that it makes memory that vital organ of our mental life whose access is indispensable for both imagination and the intellect. And once again mistakes can be made because imagination, in its attempt to recall some important details, tends to add things that are not part of the concrete object. The activities of the two faculties tend to overlap due to the intervention of imagination that tends to make up for the limited power of our memory. The latter can thus be seen as that part of the mind that can only (correctly) remember past experiences, any mistake or inaccuracy is not due to our memory failure as we usually say but rather to imagination that adds some extra-elements to past facts. Though memory seems to be overshadowed by imagination due to its seemingly passive function, it is precisely this passive and hence stable state which makes its support, as we will see further on,

---

[8] "الخيال قوة تحفظ مُثَل المحسوسات بعد الغيبوبة مجتمعة فيها" (Ibn Sīnā: 1992, p. 377); *ḥafiẓa*, which means to retain something in memory so that you no longer need the thing itself hence memorise, also involves the idea of preserving which suggests that the objects of memory are kept intact i.e. clearly distinguished from objects of imagination which in turn are distinguished from those of the intellect.

[9] ibid., p. 379.

absolutely crucial in the intentional act of generation by repetition.[10] Hence when imagination is given a total free rein, it is capable of creating bizarre things, like a phoenix[11] or a flying horse, by combining, composing and attaching in a variety of ways abstracted entities and that is why it seems to us that there is virtually no limit to what our imagination can do. In other words, according to Ibn Sīnā, imagination is also capable of performing the reverse function of abstraction. Imagination is then

[10]The role of memory in the foundation and activity of science has been little recognised by contemporary philosophers. Epistemologically, memory represents that vital region of the mind by which the epistemic subject is aware of its objectivity since its objects provide the evidence that his past experiences are not and cannot be of his own making but acquired through his constant interaction with the outside world.

[11]Ibn Sīnā has discussed in more detail the ontological and epistemic status of fictions such as phoenix (*'anqā' mughrib*) in his *Risālah fī al-nafs* or *Letter on the Soul*. He is particularly very critical of imagination and recommends it be used with great care in the sciences other than mathematics.

> Representation in imagination, except in mathematics, is often misleading (مضل), and it is a guide and heuristic (هاد و مرشد) in mathematics (Ibn Sīnā: 1956, p. 197).

> It is because the intended meaning of geometric expressions (معاني ألفاظ الهندسيات) is unequivocally known by the realisation (بالتحصيل) of its relation to its objects, and as a result no object can be imagined other than the one intended by the meaning (فلا توهم غير المعنى المقصود به). Rather each expression has an understood meaning according to the intention (بل لكل لفظ منها معنى مفهوم بحسب الغرض). (ibid., p. 196).

The active nature of imagination tends thus to confuse the mind because of the polysemy of language and its ability to create its own objects i.e. fictions such as a flying horse or a phoenix. And it seems to him that anything that can be imagined can be posed as a possible existent for possibility assumes consistency and consistency should be proved, since from imagined entities the mind can indeed be led to believe to their existence in the external world. To prevent such unwarranted and misleading inference, he requires a proof of existence from any claim that tends to confuse real with imaginative objects or any attitude that tends to substitute the latter for the former. And for the realm of natural beings instantiation seems to be the only way to prove consistency. In his *Maqāṣid* or *The Intentions of the Philosophers*, al-Ghazālī further refines Ibn Sīnā's theory of knowledge by describing how knowledge turns mental into actual existence.

> When it happens to us to imagine something we desire, from this imagination results the power of desire. If this desire is intense and perfect and our judgment that it should be is added to it, from this comes a virtue which runs through the muscles, then it moves the tendons and results in the movement of the limbs that serve us as instruments; and from there comes the desired action. Similarly, when we imagine a line that we want to draw and we judge what it should be, from the desire that that line should be comes the power to make it; the power of desire moves therefore the hand and the pen, and from them the line will result as we had imagined it. When we said that "it should be", we meant that we know or we believe that this existence will be, for us, useful, pleasant or good. The movement of the hand comes thus from the power of desire, and the movement which is the power of desire comes from the imagination and knowledge that the thing should be. We therefore find in our knowledge of a thing the beginning of its execution فقد وجدنا العلم فينا مبدأ لحصول شيء (Al-Ghazālī 1961, p. 236).

characterised by having some specific contents (signs, images, diagrams, models) by which we can refer to individuals and particulars.

4.4 The Intellect (*al-'aql*)

This is the region of understanding which enables us to have an immediate access to objects through meaning. Unlike memory which re-presents objects using some of their specific contents (signs, distinctive marks, images, diagrams), the intellect has the capacity to refer directly to them only through language. It is also in that region of the mind, and more precisely its theoretical part, which is particularly mobilised in formal sciences such as logic and pure mathematics where the mind has only to contend with itself. But why cannot memory have a direct or *absolute* access to objects like *Zayd* for example?

> As for the inner re-presentation (الخيال الباطن), it re-presents *Zayd* with those accidents and cannot abstract him absolutely (تجريده المطلق) from them. But it abstracts him from the aforementioned [positional] relation upon which sense was dependent; so his form [*Zayd*] is re-presented in memory while its bearer is absent. But the intellect (العقل) can abstract the quiddity enclosed in the individualising strange concomitants (اللواحق الغريبة المُشَخِّصة) and establish it in such a way as if it manipulates the sensible to a form of intelligible. As for that which is in its essence free from the material defects and the strange concomitants, which are not necessary to and do not follow from its quiddity, is intelligible in itself (فهو معقول لذاته), it does not need any manipulation to be prepared to be intelligible.[12] (Ibn Sīnā 1992, pp. 370–372)

An immediate access to a concrete object that is no longer present to the senses requires a complete abstraction, and this is a different mental act from the first in which the mind seems to be more passive than active. In memory or more broadly imagination, we can have only a partial abstraction of the object that is why complete abstraction cannot be done by memory because it needs a specific mental act. In both memory and imagination we gain access to an object as being this or the other, we gain access to it in some of its aspects but by intellection we gain access to the object independently of being this or the other, that is we have *absolute* access to it[13]: in fact as we will discuss later on this access amounts to grasping it as instantiating an

[12] و أما الخيال الباطن فيخيله مع تلك العوارض، لا يقدر على تجريده المطلق عنها، لكنه يجرده عن تلك العلاقة الذكورة التي تعلق بها الحس، فهو يتمثل صورته مع غيبوبة حاملها. أما العقل فيقتدر على تجريد الماهية المكنوفة باللواحق الغريبة المشخصة، مستثبتا إياها كأنه عمل بالمحسوس عملا جعله معقولا. و أما ما هو في ذاته بريء عن الشوائب المادية، واللواحق الغريبة التى لا تلزم ماهيته عن ماهيته ، فهو معقول لذاته ، ليس يحتاج إلى عمل يعمل به بعده لأن يعقله ما من شأنه أن يعقله، بل لعله من جانب ما من شأنه أن يعقله.

[13] It is remarkable that Ibn Sīnā's successors like the great scientist al-Ṭūsī for example, in his extensive comment of *al-Ishārāt*, has perfectly grasped his basic conception of knowledge by elaborating: "Intellection is the apprehension of the thing inasmuch as it is and nothing else (التعقل إدراك الشيء من حيث هو هو فقط); not inasmuch as it is another thing, either considered alone or with its other attributes that are apprehended by this kind of apprehension." (Ibn Sīnā: 1992, pp. 367–368)

invariant.[14] But this passage tells us more, the apprehension of essences also requires complete abstraction passing first through sensory perception:

> Conceptualising the intelligibles is acquired only through the intermediary of sensory perception in one way, namely that sensory perception takes the perceptible forms and presents them to the imaginative faculty, and so those forms become subjects of our speculative intellect's activity, and thus there are numerous forms there taken from the perceptible humans. The intellect, then, finds them *varying* in accidents (فيجدها العقل متخالفة بعوارض) such as it finds *Zayd* particularised by a certain colour, external appearance, ordering of the limbs and the like, while it finds *Omar* particularised by other [accidents] different from those. Thus the intellect receives these accidents, but then it extracts them, as if (كأنه) it is peeling away these accidents and setting them to one side, until it arrives to the meaning in which humans are common and in which there is *no variation* (المعنى الذي يشترك فيه و لا يختلف) and so acquires knowledge of them and conceptualises them.[15] (Ibn Sīnā 1956, p. 222; my emphasis).

The intellect can thus be seen as a second order mental act by which understanding is reached when the mind is capable of referring directly or absolutely to objects and this absoluteness is achieved by identifying a semantic invariant through its accidental variations. Better, as we will discuss in next chapter, the point is to grasp an individual as instantiating a given invariant. Hence, the concept of number is based on the identification i.e. apprehension of a semantic invariant between intentional objects,[16] and, as we shall see, is the result of a construction by iteration from unity: But how can this process be described from experience? How can an invariant whose apprehension is based on experience be conceptually expressed? And how from a particular essence like humanity can the mind conceive the general concept of number? The answers to these questions will be the subject of discussion of the next sections but let us point out that the type of abstraction involved is made of two steps at the end of which a given individual is grasped as instantiating a given concept or invariant that defines the corresponding unity. The point is that plurality and unity must logically be defined one in terms of the other using instantiation.

---

[14]The support of memory is equally necessary for the activity of pure reason.

[15]ونقول إنه إنما تكتسب تصور المعقولات بتوسط الحس على وجه واحد ، وهو أن الحس يأخذ صور المحسوسات ويسلمها الى القوة الخيالية فتصير تلك الصور موضوعات لفعل العقل النظري الذي لنا، فتكون هناك صور كثيرة مأخوذة من الناس المحسوسين، فيجدها العقل متخالفة بعوارض مثل ما نجد زيدا مختصا بلون وسحنة وهيئة أعضاء، وتجد عمرا مختصا بأخرى غيرتلك. فيقبل على هذه العوارض فينزعها فيكون كأنه يقشرهذه العوارض منها ويطرحها من جانب حتى يتوصل الى المعنى الذى يشترك فيه و لا يختلف به، فيحصلها و يتصورها. و أول ما يفتش عن الخلط الذي في الخيال فإنه يجد عوارض وذاتيات ، ومن العوارض لازمة وغيرلازمة ، فيفرد معنى معنى من الكثرة المجتمعة في الخيال ويأخذها الى ذاته.

[16]The famous mathematician, physicist and philosopher Ibn al-Haytham, who is contemporary with our philosopher, too founds science on the apprehension of an invariant by the epistemic subject:

> Science is an opinion which does not change, and the opinion is a belief in a certain notion. Science is therefore a belief in a certain notion, as it is, and it is moreover an invariable belief… But there is no belief without a believer and a believed notion." (in Rashed 2002, p. 445)

Chapter 5
The Logico-Epistemic Construction of Numbers

5.1 The Basic Epistemic Concepts of the Construction of Numbers: Aggregate, Repetition, Combinative Unity

5.1.1 Plurality as Aggregate

According to Ibn Sīnā's epistemic standpoint, essences such as humanity or white cannot be formed by abstraction from perceiving concrete things; since as he argues above, when a specific human being or a white thing is present in the mind or more precisely in memory what is perceived is this individual being *Zayd* or that white thing. Moreover, how do we come to the general notion of number? For this kind of abstraction or rather extraction is not done in the first experience, at least another experience is needed; it is not a question of time though temporality is involved. The point is that the concept of number is linked to the operation of building an aggregate:

> Plurality is the aggregate *mujtama'* of units, we have included unity in the definition of plurality, and we have done something else, we have included the aggregate mujtama' in its definition. (p. 79, § 2)

Certainly if we understand aggregate as plurality then circularity threatens again. In fact it looks as if Ibn Sīnā considers *mujtama'* or aggregate to be different from *kathra* i.e. plurality or multiplicity. While plurality, as Ibn Sīnā explicitly points out, merely means more than one,[1] the concept of aggregate or the now commonly used word "set" suggests instead not only the formation of a single and hence a specific entity from unities, but *points to a process involved for its formation*. The translator

[1]"Plurality is understood in two senses. One [the common sense] is that, with respect to the numerical ones, a thing should have that which is more than one, this being not all in relation to something else. The other [his own] is that the thing should have that which is in another thing and more. This latter is the plural in its relation to another." (p. 95, § 14).

H. Tahiri, *Mathematics and the Mind*,
SpringerBriefs in Philosophy, DOI 10.1007/978-3-319-25238-4_5

is more accurate by rendering *mujtama'* as aggregate than set for the Arabic word involves the act of collecting or putting together unities into a single amount or total as Ibn Sīnā explains in his description of the meaning of number ten (cf. supra). It is crucial to remind that, according to Ibn Sīnā, number is not a *property* of an aggregate (see Sect. 3.4, p. 35) but it emerges from a process that builds that aggregate—this makes his approach similar to the views that Frege expressed later.[2]

In fact, it very much looks as if the concept of aggregate of our author is close to the intentional concept of set of modern constructivism where sets are built by means of an inductive operation that allows to recognise if an individual is or not an element of the set, and the explicit definition of the correspondent equivalent class.[3] Let us follow Ibn Sīnā's own development. The first stage of the development points out that the elements of an aggregate are separate units, the combination of which yields a number

> The one by accident consists in saying that something united with another thing is that other and that both are one (p. 74, § 2).

The same idea is forcefully repeated in Book 1 Chapter 5:

> If you said 'The reality of A is something and the reality of B is another thing,' this would be sound, imparting knowledge; because in saying this you make the reservation within yourself that the former is something specific, differing from that other latter thing. This would be as if you said, "This is the reality of A, and the reality of B is another reality. (p. 25, § 11).

> it is like the very unity which is the principle of number—I mean, that which is such that if something else is added to it, the combination of the two becomes number (p. 76, § 11).

In each of the three examples, he describes what seems to be needed for the formation of number 2 by appealing to a thing other than the previous thing. What is needed now is a specific act of repetition:

> The only meaning of repetition one understands in this is the bringing to be of something numerically other than the first (p. 253, § 14).

If we stick to this description literarily it sounds circular: it is an operation of repetition that yields something *numerically different* to the individual to which it is

[2]Here is one of the passages in which Frege argues against the view that number is a property of things; "It marks, therefore, an important difference between colour and Number, that a colour such as blue belongs to a surface independently of any choice of ours. […] The Number 1, on the other hand, or 100 or any other Number, cannot be said to belong to the pile of playing cards in its own right, but at most to belong to it in view of the way in which we have chosen to regard it; and even then not in such a way that we can simply assign the Number to it as a predicate." (Frege 1884, p. 29, § 22).

[3]Per Martin-Löf, *Intuitionistic Type Theory—Notes by Giovanni Sambin of a series of lectures given in Padua, June 1980*, Naples, Bibliopolis, 1984.

applied. It is tempting for the modern reader to understand the idea behind as describing the operation of *successor of.* Ibn Sīnā is thus able to reduce the basic conceptual apparatus needed to define numbers to just two elements: unity and repetition or a first element and a process and this is indeed close to the notion of constructive set in *Constructive Type Theory* (Martin-Löf 1984).[4]

5.1.2 Repetition as Operator and the Role of Memory

It is important to note that, in the last quoted passage, Ibn Sīnā chooses to further emphasise his conception of repetition in a polemical context in which he dismisses out of hand "those who generate number through repetition with unity remaining constant for the one," for he argues "if repetition enacts a number and each of the first and the second does not have unity, then unity is not the principle for the composition (تأليف) of number" (p. 253, § 14).[5] His objection is that if by repetition is meant that the same unity is given to each object to be counted, then the result is unity and not number because unity does not play here the role of the principle of the composition of number. He expresses this idea by saying in the above passage

[4]Sets (as types) are defined in CTT by means of defining their *canonical elements*, those that "directly" exemplify the type and the non-canonical ones, those that can be shown using some prescribed method of transformation that they are equal (in the type) to a canonical one: the precise requirement is that the equality between objects of a type must be an equivalence relation. The implementation of such a device for the construction of a set is both insightful and straightforward:

"This we do when we say that the set of natural numbers N is defined by giving the rules:

$$0 \in N \qquad\qquad \frac{a \in N}{a' \in N}$$

by which its elements are constructed. However, the weakness of this definition is clear: 10^{10}, for instance, though not obtainable with the given rules, is clearly an element of ℕ, since we know that we can bring it to the form a' for some $a \in N$. We thus have to distinguish the elements which have a form by which we can directly see that they are the result of one of the rules, and call them canonical, from all other elements, which we will call noncanonical." (Martin-Löf 1984, p. 7).

[5]This is strikingly similar to Frege's objection that will be discussed later in detail in the next section.

that the second does not, that is, it cannot be said to have a real unity. In other words, this conception of repetition confuses the essential (or metaphysical) unity which could be seen as a property of every single object and the numerical unity as the principle of the composition of number.[6] Repetition here fails to function as an

---

[6]In his *Commentary on* al-Ilahiyāt, al-Shirāzī better known as Mulla Ṣadrā (1572–1640) further illuminates what Ibn Sīnā has in mind by this distinction: "the unity which is a principle of the mathematical numbers is other than the unity which can be found in the separate substances for the separate substances do not possess a quantitative number generated by repetition of similar units." (in Avicenne 1978, p. 302) Mulla Ṣadrā's comment indicates that it is Ibn Rushd, generally known for his hard-line Aristotelianism, who misinterprets Ibn Sīnā by accusing him of confusing the numerical and the metaphysical unity. In the third discussion of his *Tahāfut al-tahāfut*, he blames the contradictions raised by al-Ghazālī on his predecessors' innovations.

> All these are inventions fabricated against the philosophers by Ibn Sīnā and al-Fārābī and others. But the true theory of the ancient philosophers is… (Ibn Rushd 1930, p. 184)

His project is to show that Aristotle's system is free from contradiction, but he also assumes that it can succeed in explaining what motivates his predecessors' innovations. It is in this context that he presents us his *Talkhīṣ mā ba'da al-ṭabī'a* known as *Epitome of the Metaphysics* as an example to show us how Ibn Sīnā has erred by fundamentally confusing the numerical and metaphysical unity. And in the course of his polemical discussion, the Commentator was led to confront some of the new questions posed by the development of the concept of number. We here summarise some of his main views. Like the author of *al-Shifā'*, he considers presence in the mind as mental existence, his favourite examples are goat-stag and the void.

> Goat-stag and the concept of void and other similar things that the mind composes have no existence outside the soul in the manner of that composition. (Ibn Rushd 1958, p. 57)

More generally:

> The intelligibles which exist only in the mind, this is something which is not impossible (المعقولات التي وجودها في الذهن فقط، فذلك ليس يمتنع) for this meaning is one of what we enumerated to which the term existence applies. However, this meaning and the meaning by which this [term] signifies the essences individually are entirely distinct. All this becomes simply clear upon a moment in reflection, but this is the case of this man [i.e. Ibn Sīnā] who adds many things from his own mind. (ibid., pp. 10–11)

Like Ibn Sīnā, he argues that the existence of numbers is due to a specific activity of the mind and recognises that their mental existence is of different order than that of geometric objects.

> The numerical one is the meaning of the individual abstracted from quantity and quality, I mean that by which the individual is an individual. Since it is only an individual in the sense of being indivisible, then the mind abstracts it from matters (فيجرده الذهن من المواد) and grasps it as a separate meaning. For the one in number and numerical unity are only something which the soul produces in existing individuals (إنما هو شيء تفعله النفس في أشخاص الموجودات); were it not for the soul, there would exist neither numerical unity nor number at all unlike the lines and the planes and more generally continuous quantity. That is why number is the more abstracted from matter. (ibid., p. 103).

As a result, Ibn Rushd admits that numerical unity is more general than metaphysical unity that he identified with the individual, i.e. the single object that actually exists outside the mind.

operator since its object is not the previous result or in Ibn Sīnā's words unity is not applied as a result of linking the current to the previous experience. And he immediately repeats again what is needed to make a repetition an operator: "if one inasmuch as it is one is unity and the second inasmuch as it is second is unity, then there are two unities" (p. 253, § 14). The number 2 cannot be generated by assigning the same unity to the second object to be counted but by applying another unity, i.e. a unity other than the first, and as he concludes before this is "the only meaning of repetition one understands". Ibn Sīnā seems to present his conception of repetition as corresponding to the underlying counting process performed by an

---

(Footnote 6 continued)

> In this discipline [metaphysics] the one is used as equivalent to essence of the thing and its quiddity, among which one in number that can be used to signify the individual which cannot be divided as an individual, like our saying: one man and one horse. (ibid., p. 18)

However, instead of equating metaphysics with the extended ontology like the author of *al-Ilāhiyāt* who warns us that his metaphysics "should be learned after the natural sciences and mathematics" respectively (Ibn Sīnā 2005, p. 14 § 6), the author of *Talkhīṣ* chooses to restrict the scope of his own metaphysics by excluding mathematics as one of its subjects as he stated in the introduction: "as for its order in education, it [metaphysics] should be after the science of nature." (Ibn Rushd 1958, p. 7). The above and other passages dealing with the nature of mathematical objects show that the Commentator has abandoned the Stagirite's mathematical doctrine. For the conception of number, that he was led to defend and argue for, is not the one that comes from his comments on Aristotle's *Metaphysics* whose conception of number is not referred to, but from his predecessors' innovative ideas. In his effort to fight his opponents' innovations on their own terrain, he has unwittingly ended up adopting some of them vindicating thus their departure from Aristotelian philosophy. *Talkhīṣ* is a misleading title since the meaning of the Arabic word suggests that it is one of the author's short commentaries on Aristotle's *Metaphysics*. Its content was only clarified after the edition and publication of two major works of al-Fārābī, *Kitāb al-ḥurūf* (*The Book of Letters*) and *Kitāb al-wāḥid wa al-waḥda* (*The Book on the One and the Unity*), for it turns out that in at least two out of the four chapters, Ibn Rushd was heavily using al-Fārābī original ideas without ever mentioning him by name in the entire book. It is perhaps curious that he is not after all that loyal follower of Aristotle on this fundamental chapter of theoretical philosophy as he wished to be. But this is by no means surprising. For this peculiar feature of theory change is a recurrent phenomenon in the history of science and thus familiar to the historians of science: when an old theory is used to explain new scientific facts for which it was not originally designed to account for, scientists tend to invent new concepts that breaks with the old theory. This is what happens to Ibn Rushd's predecessors as pointed out by Rashed. What is somewhat surprising is that this process of change is not recognised by scholars and historians of philosophy despite the clear evidence provided by the main actors themselves. Ibn Sīnā has explicitly announced his break with the Aristotelian tradition in his *Logic of the Orientals*. And while in his introduction to the work, he seems to think that his reforms came too little too late, Ibn Rushd's strong reaction suggests that they went too far too fast. He was well aware of Ibn Sīnā's major innovations and he blamed them for throwing philosophy into disarray. But Ibn Rushd appears to accept the emergence of new epistemic concepts like production of the mind, mental existence, *tajrīd* or abstraction, which are all as it happens Avicennan concepts, and he seems to have taken them for granted as he makes little attempt to discuss them. This prevents him from recognising that these are not just scattered and arbitrary words that appeared by chance, they are rather part of a new epistemic framework forged by the physician to account for the formalisation of mathematical ontology.

epistemic agent in a phenomenological setting in which repetition is described as the result of a specific intentional act. The motivation of his phenomenological analysis seems to be to elucidate how the specific repetition operation can be actually performed by a human being (or perhaps more generally by an epistemic agent)—different to say from Peano-Dedekind approach to the notion of successor or Frege's concept of hereditary chains. This step might be motivated by Ibn Sīnā's interest in the actual practice of mathematics, thus a proper epistemology characterising the specific act by the means of which numbers are generated is due. The point of the discussion seems to explain his understanding of repetition which consists in applying another unity to the previous one, i.e. why the formation of a number, like 2 for example, requires a unity other than the first. His explanation could also be regarded as an attempt to clarify for those who confuse the numerical unity, which is the principle of the composition of number, with the essential unity by showing how the meaning of the first emerges from the latter.

The departure point of the phenomenological understanding of the repetition act is that the new and the former experience are linked with the help of memory or more broadly, imagination. Imagination re-presents the first experience stored in memory, i.e. the first intentional object which is no longer here, the imagination acts thus as a reminder (*munabbih*): the present white thing acts as a reminder "*munabbih*" of the previous white thing. That is why the intellect apprehends forms through memory. If, due to some fatal mental disease or major mental disorder, for example, imagination fails to re-present the previous experience required for the apprehension of the invariant by the intellect, then there is no *munabbih* i.e. no reminder and consequently the number of the second experience will not be experienced as a second one but as a new one.[7] A series of different experiences is useless if each one cannot be linked to the previous one. Likewise, no new number can be generated if an intentional object is or perceived as exactly the same as the former (two simultaneously given identical objects are nevertheless differentiated by their localisation). This is what prevents Ibn Sīnā from defining from the outset number in terms of repetition for the kind of repetition required is not the commonly understood i.e. the same thing or the same unity but the *different same* that can only brought about by another experience. That's why the generation of a number, such as 2 for example, out of a previous one requires the emergence of the new experience of an act that should be linked to the previous one in such a way that the new experience is apprehended and re-cognised as composed by the repetition of the first—again in this example it seems to be very natural to understand this as describing the phenomenological act that corresponds to the *successor* of 1:

[7]For the physician, the various faculties of the mind that he described are not simply conceptual instruments designed to explain the possibility and activity of science. They correspond to concrete regions that can be localized in the brain, the reason is very simple "what leads people to judge that these are the organs [of the faculties] is that the corruption, when it particularly affects a cavity, results in disability." (Ibn Sīnā 1992, p. 385).

> For unity is not repeated unless it would be there in one succession after another. And this succession would have to be either temporal or essential [i.e. *de re*] (ذاتية). If it is temporal and unity is not annihilated in the intermediary stage, then it would be as it was before, not something that has been repeated. If it ceases to exist and is then brought into being, then what has been brought into being is another individuality. If the succession is essential, then this consequence is more evident. (p. 254, § 14)

By grasping repetition as an operator in which time plays little role, Ibn Sīnā explains why it is such a basic and powerful pure mathematical operation in the construction of mathematical concepts. The author of *al-Shifā'* would no doubt agree that repetition is performed in time, but time appears to be external or not *sufficient* to counting as such, for it is not essential to the bringing about of the actual occurrence of another thing let alone linking it to the previous one as a second instance. For if it were the case, time would appear as the cause of the occurrence of the same thing and thus of its mental existence.[8] This is similar to repetition in experimental sciences where repeating the same experience is also performed in time, but the reason for conducting a second experiment is to establish the regularity of the phenomena by confirming the first result. This explains why Ibn Sīnā insists so much on the actual occurrence of a second experience (yet to be re-cognized as such). For the formation of essences requires change in the mind, and change in the mind cannot happen miraculously by itself by some kind of magic but should be brought about by change in intentional objects and we shall see shortly why change is critical in the formation of numbers, otherwise his account would be deemed merely psychological.

5.1.3 Intentionality as Combinative Unity

Ibn Sīnā's epistemology embodies the idea that science is based on the critical role played by the mind in our life by linking and unifying our experiences. Moreover, his general epistemology conceives two different acts, namely what we would call a synthetic one by means of which new entities are generated and an analytic one by means of which a given unity is analysed into its elements. Accordingly, the synthetic process involves the process of linking different unities into one while the second involves the process of separating the components of a given unity. Take the example of the number 2: either we are in a context where a unit is given and then the number 2 is generated by repetition of that unit; or the number 2 is given and then we start the generation process backwards. Once more here the examples in the explanation of Ibn Sīnā seem to deal with the case where we have already different units:

> There are those [types of unity] where what is understood by them is not divisible in the mind, to say nothing about material, spatial, or temporal divisibility. […] Water is

[8]But time seems to be essential not to counting as such but to the actual recognition of identity for it looks as if the mind needs sometimes to be able to effectively objectify a phenomenon. Further discussion of this topic is out of the scope of this paper.

> numerically one in being water but has it in its potentiality to become numerically many waters-not, however, by reason of ‘waterness’ (المائية), but because of its connection with the cause: namely measure. These numerically multiple waters would then be one in species and also one in subject, because it is of the nature of its subject to unite in actuality as one in number. But this is not the case with individual people. For it is not by virtue of a number of subjects among them that they become united as the subject of one human. Yes, each one of them is one in terms of its one subject, but what accrues of the multiplicity is not one in subject; nor is its state the state of each particle of water. For the latter is one in itself through its subject. The aggregate of water particles is said to be one in subject, since it is the function of such particles to unite as one subject through conjoining, whereby its aggregate then becomes one water.[9] (p. 77, § 12–14)

Ibn Sīnā calls *combinative unity* (الوحدة الإجتماعية) the mental act that combines what is otherwise separated by nature.

> Combinative unity has plurality in act. And as a result there is plurality over which unity is superimposed, but which does not remove plurality from it.
>
> و الوحدة الإجتماعية فيها كثرة بالفعل. فهناك كثرة غشيتها وحدة لا تزيل عنها الكثرية. (p. 75, § 7)

Number is not just a name that refers to plural things like collective nouns such as people for example or more generally plural nouns such as beasts, it is rather a plurality in which discrete unities are unified by the combinative act which makes it in turn to be treated in itself as a unity, an object to which some properties can be attributed.[10] But Ibn Sīnā points out that number, as general concept, is not a thing since it is still undetermined, no particular number has yet been yielded.

> Number is not a realised thing (شيئا محصلا) so long as it has not become varied in species such as two, three or four. Moreover, if it becomes realised, its realisation does not come about through the addition to it of some external thing where there would be, besides the generic nature—such as “measureness” (المقدار) and “being numeral” (العددية)—a subsisting indicated nature to which another nature is added and through which [the generic nature] becomes varied in species. Rather, the nature of “twoness” itself is the characteristic of “being numeral” that is predicated of “twoness” and is proper to it بل تكون طبيعة الإثنينية نفسها هي العددية التي تحمل على الإثنينية و تختص بها (p. 55, § 24)

[9]Ibn Sīnā introduces here oneness or unity of substance such as one human whose essential property, as he explains, is its indivisibility in actuality. And at the end of this chapter, he concludes that the one and the existent are coextensive: “The one may correspond with the existent in that the one, like the existent, is said of each one of the categories. But the meaning of the two differs, as you have known. They agree in that neither of them designates the substance of any one thing.” An idea that he clarifies further in book 7 chapter 1, “everything has one existence. For this reason, it is perhaps thought that what is understood by both is one and the same. But this not the case, rather, the two are rather one in subject-that is, whatever is described by one is described by the other.” That’s why he makes of oneness in the next chapter 3, as we shall see, one of the first principles of conception or an a priori concept.

[10]Ibn Sīnā insists once more on the unity aspect of number that distinguishes it from collective and plural nouns: “If someone says, “plurality can be composed of things other than unities, as in the case of people and beasts,” we say: just as these things are not unities but things constituting the subject of unities, in a similar way also they do not constitute plurality but only the subjects of plurality; and moreover just as these things are ones, not unities, similarly they are plural not pluralities.” (p. 80, § 6 and 7).

As a result, particular combinative unities yield the general concept of particular numbers which holds an intermediary level between particular numbers and the general concept of number. Twoness is therefore the specific combinative unity that yields two, likewise Ibn Sīnā explicitly speaks elsewhere of "threeness", "fiveness", "tenness".[11] But how do we come to the idea of the specific combinative unities? Can we account for their formation? And if so, by what means? Ibn Sīnā did not raise explicitly these questions but they can be answered by his epistemology for the formation of the specific combinative unities remarkably follows the same pattern as the one involved in the formation of particular numbers. It is thus sufficient to repeat the same epistemic process used in the latter to account for the former.[12]

Ibn Sīnā defines twoness as "the first number, being the limit in numerical fewness"[13] (p. 94, § 11) in other words, twoness is the first combinative unity i.e. the least plurality greater than one. Let us find out its exact meaning, and how it is formed. We have already explained how From *Zayd* and *Amr*, the particular number two humans is formed; and similarly from white cat and black cat (to take Frege's example to which we will come back later) another particular number two cats is formed,[14] and something more! for according to Ibn Sīnā's epistemology, the

---

[11]Ibn Sīnā's tendency to use abstract names like "measureness" (المقدار), "being numeral" (العددية) and "twoness" (الإثنينية), etc., is particularly noteworthy. Unlike in Arabic, some of these names sound strange in English that's why the translator puts them in quotation marks. The frequent use of that type of abstract names, which are called *maṣdar* by the grammarians, and the recognition of the importance of their meaning indicate how Ibn Sīnā fruitfully exploited the resources of the root-based Arabic formation of words in his epistemology.

[12]Two important remarks about Ibn Sīnā's basic epistemic framework. The first is that it is a process which can be repeated, an iteration feature that ensures or rather explains the objectivity of his conception. The second is that Ibn Sīnā seems to consider it as one of the major principles of knowledge since, besides mathematics, he surprisingly appeals to the same basic process in his book of *Physics* to account for the apprehension of motion. To put it very briefly, for Ibn Sīnā local motion is perceived as whole and characterised by its point of departure and its finishing point where the object comes to rest; hence the possibility of the perception of motion requires, like counting for that matter, the retention of its initial past state i.e. its point of departure and the recognition that it is the same object which was moving throughout (Avicenna 2009, Book 2 Chapter 1, p. 112); further discussion of this important topic is out of the scope of this study.

[13]Al-Fārābī expresses the same idea: "The subject of one which is said of many is necessarily many, the one cannot be true of it if it is not many… and the least of that [many] is two." (Al-Fārābī 1989, p. 71, § 50).

[14]It is interesting to point out that al-Fārābī sketches similar reasoning by discussing the formation of class of classes, he captures this idea by distinguishing between one in number in which a class is considered as element and the yielded aggregate that he calls many in number i.e. class of classes. He provides some examples and one of them is the formation of classes generated from one in species, he explains that there are two ways to generate a class of classes:

> If we take for example two individuals from a species, two individuals from another species and two individuals from a third, each of the two individuals are one in species and the many generated from the three ones is many that is generated by one in species [taken as element]. But if we compare two of one species to two of another species, their aggregate is also many in species [i.e. two species] in respect to one in species [as the property that generates the plurality]. (Al-Fārābī 1989, p. 64, § 38)

particular number two cats triggers the epistemic construction by acting as a reminder for the previous particular number two humans due to the apprehension of a second order invariant that he calls one by number or similarity in number.[15] As a result twoness, the specific combinative unity that yields the general concept two, is not just {*Zayd*, *Amr*} or {white cat, black cat} but {*Zayd*, *Amr*}, {white cat, black cat} and all similar classes or more precisely the class of all classes similar to {*Zayd*, *Amr*}. From this definition of two, the more general concept of number can immediately be inferred as the class of all classes similar to a given class. Ibn Sīnā did not state this definition in such terms, but, in Book 7 in which he rejects the definitions of his opponents, he formulates his condensed conception of number[16]:

$$\text{Number is formed from similar unities (p. 252, §11)} \tag{5.1}$$

Given Ibn Sīnā's epistemic framework, this concise definition is equivalent to the just aforementioned Fregean definition of number as the class of all classes similar to a given class. Since we know that for him number is just another name for aggregate or class and formed from concrete things, (5.1) thus becomes:

Number is the class of unities similar to given unities

But unities are not only individuals, they can also be classes as we have just seen; by performing the right substitutions, the above formulation becomes:

Number is the class of classes similar to a given class

Ibn Sīnā's epistemology distinguishes thus three levels in the construction of numbers set out in an increasing order according to their generality and abstractness: particular numbers (like two humans for example), the general concept of particular numbers (the concept of two or twoness) and the general concept of

---

(Footnote 14 continued)

He then points out that the second method can be used to generate a class by choosing one individual from each species.

> But if we take an individual from each species, the many generated is in respect to one in species [i.e. individuals representing each species] and not by one in species [since the elements are individuals and not classes]. (ibid.)

[15]Similarity in number is the relation that we call nowadays one to one correspondence. The concept of similarity turns out to be very fruitful for it can furthermore be used to define the equality of numbers: two numbers are equal if they are similar or in one-to-one correspondence i.e. if each unit of one number corresponds to a unit in the other number.

[16]This new definition of number did not escape the attention of Ibn Sīnā's successors such as al-Shirāzī; in his commentary on *al-Ilāhiyāt*, he exactly repeats the author of *al-Shifā*'s definition when he says: "the number is composed of similar units." (in Avicenne 1978, p. 302).

number or "being numeral". The latter is inferred from the previous level which is obtained by repeating the same epistemic process applied to reach the first. Our author's task is then to explain the basic act involved in the formation of the three kinds of concept of number, the combinative unity. And it is surprising to find that the logical epistemic nature of his answer would meet, as though in anticipation, the very same objections raised to this kind of approaches to the foundation of arithmetic by Frege, the founder of modern logic.

5.2 The Logical Construction of Numbers: Instantiation, Invariance, Equivalence Classes

5.2.1 The Apprehension of Essence as Invariant

Ibn Sīnā tackles the issue of how pluralities can be conceived as a unique object: unities are linked into a single mental act—namely, the act of combinative unity—that provides its existence. The achieved unity is thus an intentional object ontologically dependent on the epistemic subject without whom the unity would collapse. Still, how is it that a specific single entity can be formed by linking a series of experiences? Certainly, we have the repetition act but first we have to have the units on which the repetition act can operate. How can for example the concept of *two humans* be generated by just linking an experience of *Zayd* to that of *'Amr*? The problem is that if we just apprehend the essence of both individuals we will only come to the result that both individuals fall under the concept human and we will not be able to distinguish them as differentiate units. This is precisely Frege's objection against such kind of approaches:

> For suppose that we do, as Thomae demands, 'abstract from the peculiarities of the individual members of a set of things', or 'disregard, in considering separate things, those characteristics which serve to distinguish them'. In that event we are not left, as Lipschitz maintains, with "the concept of the number of the things considered"; what we get is rather a general concept under which the things considered fall. The things themselves do not in the process lose any of their special characteristics. For example, if I, in considering a white cat and a black cat, disregard the properties which serve to distinguish them, then I get presumably the concept "cat" [not two cats]. (Frege 1884, p. 45, § 34)

Frege's objection is particularly relevant to the phenomenological approach for the difficult task is to account for identity since experiences, which are based on the perception of concrete objects, are by nature different from each other. Frege's objection is an interesting move since he seems to easily concede to his opponents that identity is grasped only to formulate the problem that it poses: how to account for or rather recover plurality once the differences have been disregarded? The target of his attack is once again the notion of abstraction of the English empiricists prevailing in his time. But this kind of abstraction is irrelevant to Ibn Sīnā's epistemology which makes all the difference. To meet the author of *the*

Foundations of Arithmetic's challenge, the author of *al-Shifā*' has to innovate. Let us find out how our author would deal with a formidable opponent like Frege.

Ibn Sīnā's first step is to argue that plurality and unity must be understood simultaneously:

> And as for plurality, it is necessarily defined in terms of the one because the one is the principle of plurality, the existence, and quiddity of the latter deriving from it. (p. 79, § 2).
>
> Here, unity is taken as conceived in itself and as one of the first principles of conception. On the other hand, our explaining unity in terms of plurality would be a directing of attention (*tanbīhan*) wherein the imaginative course is used to hint at an intelligible which we already have but which we do not conceive to be present in the mind. (p. 80, § 4)

The idea behind this assertion once more seems to come close to the approach of modern Constructive Type Theory (CTT) and other constructivists of the Erlangen-School:

> In the world being disclosed to us all along through language we tend to grasp the individual object as individual at the same time that we grasp it *as specimen of*... (Kamlah and Lorenzen 1984, p. 37)

If our interpretation is correct then the idea behind the paragraph of Ibn Sīnā quoted before is that defining pluralities in terms of unities should be understood as the act of grasping that they are instantiations of a concept (or unity) already given in the intellect—the pluralities are not given but only grasped as instantiations—and defining unity in terms of pluralities given by sensation or imagination amounts to the act of grasping that those individuals instantiate a unity not yet explicitly *present in the mind*.

We are now able to discuss Frege's example. The author of *The Foundations of Arithmetic* does not spell out how we get at the concept cat from a white cat and black cat. But since abstraction is the process to be used by "disregarding the properties which serve to distinguish them", his line of reasoning seems to run as follows:

From a white cat, by abstraction we get at the concept cat
From a black cat, by abstraction we get at the concept cat

The overall result is the concept cat, as he concludes: "The concept 'cat', no doubt, which we have arrived at by abstraction no longer includes the special characteristics of either, but of it, for just that reason, there is only one." (Frege 1884, pp. 45–46, § 34).

Frege blames abstraction for failing to get at the concept of number without telling us exactly why. For we can obtain the same result without using abstraction:

A white cat is an instantiation of the concept cat
A black cat is an instantiation of the concept cat

The overall result is the concept cat. Both examples are part of the general conception, already discussed (Sect. 5.1.2, p. 53), in which number is generated by just repeating the same unity. They differ in the reason provided for appealing to such repetition.

For Ibn Sīnā, the two examples would only show that it is Frege's analysis which is misleading by putting on equal footing instantiation and abstraction. His phenomenological description leads him to retain the first and ignore the second:

From a white cat, we get at a white cat
From a black cat, we get at a black cat, and something more: the concept cat.

His epistemology is precisely worked out to account for this fact: the concept cat is reached by an epistemic mental construction triggered by the black cat acting as a reminder for the previous experience, that of white cat, due to the identification of an invariant. The black cat and the white cat are the same with respect to "catness," but they are different with respect to each other since for Ibn Sīnā the mind does not thereby remove the particularities from concrete objects. He ignores abstraction because the mind cannot apprehend the concept cat from just the perception of a (single) concrete object. And as if to effectively meet Frege's challenge, instantiation is used to put his phenomenological analysis into a rigorous formulation: when I grasp the black cat I grasp it as instantiating a concept as an invariant that provides the unity and similarity for the white cat. In other words the white cat and the black cat are grasped (with their own differences) as different instantiations of the concept cat.

Thus Ibn Sīnā conceives of instantiation as a one-to-many relation (next section) not a many-to-one relation as generally believed, an analysis that challenges the commonly used expressions such as instantiation of or falling under a concept whose ordinary meaning conceals the fact that a concept can only be grasped as an invariant which necessarily requires a new experience. Like abstraction, the flaw in the instantiation example is precisely the belief that we can distinguish an object and a concept by just perceiving one and the same object. While for Ibn Sīnā, the identification or rather the re-cognition of an invariant is due to the occurrence of another experience, in this case the black cat, which makes us aware that both experiences, the current and the previous i.e. the black cat and the white cat, are in fact instantiations of the same concept cat, i.e. that they are different elements of the same equivalence class. So what is required now is the notion of being equal in an aggregate, or more precisely, we need a notion of equivalence class. This is a necessary condition for the construction of an aggregate—for further discussion of this see next section.

But what about the aggregate of numbers itself? Well, they are generated by some kind of inductive definition as described in the preceding section. However, the point is that performing this inductive definition by an epistemic subject is made possible by the act of remembering the grasping of the first individual, then remembering the grasping of grasping the first individual and so on:

> It is within the power of the soul to apprehend intellectually, and [once it has made its first apprehension] to apprehend that it has apprehended, and to apprehend that it has apprehended that it has apprehended, and to construct relations within relations and to make for the one thing different states of relations ad infinitum in potency.[17] (p. 160, § 8)

[17] إن في قوة النفس أن تعقل، و تعقل أنها عقلت، و تعقل أنها عقلت أنها عقلت، و أن تركب إضافات في إضافات، و تجعل للشيء الواحد أحوالا مختلفة من المناسبات إلى غير النهاية، بالقوة.

> For there are, here, relationships in irrational roots and in numerical relations that are easily accessible (قريبة المنال) to the soul, and the soul does not need in one state to ([actually]) intellectually apprehend all of these or to be constantly preoccupied with them. Rather, it is within its proximate power (إن في قوتها القريبة) to intellectually apprehend this—as, for example, bringing to mind infinite multiple progressions, bringing to mind the infinite doubling of numbers and, indeed, the infinite recurrence (مرارا لا نهاية) through doubling of the same relation between a number and one similar to it (مثله).[18] (p. 161, § 8)

Like al-Khwārizmī before him (next section), Ibn Sīnā finds in instantiation, like intentionality and invariance for that matter, such a powerful logical concept that it captures the relation between concrete objects and the concept to which they give rise making superfluous any further philosophical analysis; and it turns out to be so rich that it gives rise to rules and procedures which govern its use making futile any attempt to look for a supposedly (more) intuitive description of the process. It is remarkable that Ibn Sīnā has anticipated a challenge formulated later by Frege, and it is even more remarkable that he surprisingly meets this challenge by providing a logical answer. For the crux of Ibn Sīnā's answer to Frege's challenge is not that there is a process by means of which we extract an invariant but rather that there is no other way to grasp an individual other than as an instance of the invariant that defines the corresponding unity. From Ibn Sīnā's epistemic perspective, invariance/instantiation are correlative terms that explain what seems to be the paradoxical feature of number as unity in plurality. Moreover the difference within a given equivalence aggregate is maintained for the elements do not and need not lose their specific differences for its constitution. More importantly, Ibn Sīnā's analysis can be seen as *partially* vindicating Frege's view by unexpectedly showing how logic is, after all, essentially part of the foundations of mathematics and contribute to the understanding of its evolution, and his inclusive approach illustrates how the dynamic interaction between mathematics, logic, epistemology, phenomenology, linguistics and law contribute to their own development.

Another kind of objection is the following: this account which is based on the apprehension of an invariant cannot explain the formation of numbers from concrete different objects which have no species in common like for example *Zayd*, Bagdad, white. Ibn Sīnā's answer would be: if totally different objects can be accounted at all, it is only by virtue of the apprehension of an invariant such as *to be a thing*. In fact, this is a consequence of defining pluralities as instantiating a given unity. Ibn Sīnā's epistemology is symptomatic of the major change brought about by al-Khwārizmī's *Kitāb al-Jabr* that drives him to establish that it is in the nature of the mind to go beyond its intentional object by indefinitely relating it to others in a variety of ways, and recurrence, i.e. repetition of the same epistemic mental act, turns out to be just one of the simplest relations, for the powerful creativity of the human mind can construct more complex relations by reasoning by induction over relations. It is remarkable that the 11th century Islamic physician is involved in

---

[18] فإن ههنا مناسبات في الجذور الصم و في إضافات الأعداد كلها قريبة المنال من النفس، و ليس يلزم أن تكون النفس في حال واحدة تعقل كلها أو أن تكون مشتغلة على الدوام بذلك، بل في قوتها القريبة أن تعقل ذلك مثل إخطار المضلعات التي لا نهاية لها بالبال، و مزاوجة عدد بأعداد لا نهاية لها بالبال، بل بوقوع مناسبة عدد مع مثله مرارا لا نهاية لها بالتضعيف.

grounding the notion of mathematical induction in a way that resembles so much the one of the 19th century French mathematician Henri Poincaré:

> It [i.e. mathematical induction] is only the affirmation of the power of the mind which knows it can conceive of the indefinite repetition of the same act, when the act is once possible. (Poincaré 1902, p. 41)

In the following section further support will be given for the relation between plurality and unity in the context of Ibn Sīnā's notion of equivalence classes or aggregates.

5.2.2 *Essence as Equivalence Relation*

Another major innovation of Ibn Sīnā's mathematical investigations is his attempt to formalise his own conceptual analysis of the formation of numbers within equivalence classes. We have already examined how he expressed his condensed conception of number in terms of similarity:

Number is formed from similar unities and nothing else. (p. 252, § 11)

i.e. the idea of what we called the *different same* is here captured by the more functional similarity relation. This is confirmed by what Ibn Sīnā tells us elsewhere when he discusses the different means by which unity of different objects is achieved.

> Unity is similar, whereas what is contrary to it [plurality] is varied, changeable, and ramified. […] That which is the same in quality is the similar, that which is the same in quantity is the equal, that which is the same in relation is called the corresponding… as for the same in essence, what is the same in species is called similar that which is called similar in properties is called resembling. (p. 237, § 2)

Ibn Sīnā's underlying idea is to systematically turn predicates like to "be human" into "similar in humanity", to "be animal" into "similar in animality"; "white" into "similar in colour" or "governance" into "similar in state of affairs". But why make such conversion? And what do we gain by making it? It is in Book 5 Chapter 1 where Ibn Sīnā reminds us of the meaning of abstracted entities like humanity:

> Universality occurs to some nature if such nature comes to exist in mental conception (التصور الذهني). […] That aspect of "the human" that is intellectually apprehended in the soul is the universal. Its universality [however] is not due to it being in the soul [i.e. psychologically conceived] but due to it relating (مقيس)[19] to many individuals [i.e. logically

[19]This is remarkably the same word used by al-Khwārizmī in his landmark book *Algebra and al-Muqābala* to capture the relationship between particular equations and the general equation of which they appear as an instantiation. Ibn Sīnā describes the relation of a universal to individuals as one-to-many, he further makes sure to distinguish it from the psychological conception making it in effect a logical epistemic relation, that's why we call logical epistemic his construction of the natural numbers. We have seen that this is the relation that he captured by "falling under a concept" in his later book *Pointers and Remainders*.

> related], existent or imagined, that are governed by it by the same governing rule [of instantiation].[20] (p. 159, § 5)

This is clearly linked to the view discussed in the preceding sections that humanity has no meaning other than its instantiation by individuals. For Ibn Sīnā then essences cannot really be defined but only subject to instantiation for we can say of each object presented to us whether it is a human or not. And since the apprehension of essences requires at least two objects, predicates become de facto part of the class of relations. As a result essences can be defined in terms of the conditions which should be satisfied by the relation between objects. And what is the main property involved in effectively constructing an equivalence class? Ibn Sīnā tells us in the case of humanity what does it mean for example for an individual like *Zayd* to be similar to another individual like *'Amr*?

> If the humanity in *'Amr* (taken as entity by itself, not in the sense of a definition) exists in *Zayd* then whatever occurs to this humanity in *Zayd* would necessarily occur to it when in *'Amr*. (p. 158, § 4)

Indeed one of the consequences of being equal members of an equivalence class is that substitution applies (and substitution is made possible by the symmetry; reflexivity and transitivity of the relation). Three examples have been identified so far in which Ibn Sīnā effectively applies substitution: the first example can be found in the above quoted passage (Sect. 3.4, p. 35) in which Ibn Sīnā performs the substitution to prove the circularity of Euclid's definition.

Example 1 (see Sect. 3.4, p. 35)

Number is a plurality composed of units (Euclid's definition); but
Plurality is composed of units, and
Plurality is the same as number ("العدد نفس الكثرة")[21]; hence
Plurality is plurality (respectively, number is a number)

The other two examples in which the substitution is used are identified by Hodges and Moktefi and also by Rahman and Salloum (Rahman and Salloum 2013, p. 55)

---

[20] تعرض الكلية لطبيعة ما إذا وقعت في التصور الذهني... فالمعقول في النفس من الإنسان هو الذي هو كلي، و كليته لا لأجل أنه في النفس، بل لأجل أنه مقيس إلى أعيان كثيرة موجودة أو متوهمة حكمها عنده حكم واحد.

[21] Ibn Sīnā explicitly states here an equality relation between the two members, hence we can substitute whatever term we like in the first proposition. A similar proof of a circular definition using substitution is expressed in the following passage:

> If you say 'The thing is that about which (ما) it is valid to give an informative statement الشيء هو ما يصح الخبر عنه,' it is as if you have said, 'The thing is the thing about which (الذي) it is valid to give an informative statement الشيء هو الشيء الذي يصح الخبر عنه' because the meaning of 'whatever' (ما), 'that which' (الذي) and the 'thing' (الشيء) is one and the same. You would have then included 'the thing' in the definition of 'the thing'. (p. 24, § 7).

Example 2 (al-Qiyās 472.15f)[22]

Zayd is this person sitting down, and
This person sitting down is white.
So *Zayd* is white.

*Example 3 (*al-Qiyās *488.10)*[23]*:*

Pleasure is B.
B is the good.
Therefore pleasure is the good.

For examples of the transitivity relation[24]:

If A is equal to B and B is equal to C, then A is equal to C[25]

Ibn Sīnā also states (and even attempts to prove) the transitivity of the parallel relation in his *Danesh Name* or *The Book of Scientific Knowledge*: "We say when a line is parallel to a second line which is parallel to a third, it follows that the first is parallel to the third" (Avicenne 1986, II p. 97), i.e.

$$\text{If } L_1 \parallel L_2 \text{ and } L_2 \parallel L_3 \text{ then } L_1 \parallel L_3$$

The problem might now be that if the underlying equivalence classes are to be taken as the essence of the individuals that instantiate that class, what about the accidental properties? Perhaps one can see them as the predicates that are defined on the equivalence class. Such a strategy would assume that we distinguish two different predication acts that correspond to two different types: the underlying set and the propositional functions built on that set. The predication *a is B* is ambiguous then, it might be understood as *a instantiates the aggregate B* (*a is an element of the set B*) and as the proposition *Ba is true* where *a* instantiates the aggregate *A* (*Ba is true and a is an element of the set A*) (see Rahman et al. 2014, p. 298). In his discussion of a wide range of examples of building equivalence classes in Book 3 Chapter 2, Ibn Sīnā presents us with the following case (for a formal study of this and similar cases see Rahman et al. 2014, p. 305):

[22]Hodges and Moktefi (2013, p. 79).

[23]Ibid. p. 95.

[24]It is interesting to point out that al-Jurjānī (1340–1413) knows the transitivity relation that he calls the rule of equivalence (*qiyās al-musāwāt*). In his famous *Mu'jam al-ta'rīfāt* or *Dictionary of Definitions*, he provides two kinds of example to illustrate his definition. The first represents the class of relations which are transitive like equality, and the second those relations which are not transitive like "to be half (*niṣf*)" as he explains: "A is half of B and B is half of C, it is not true to infer (*falā yaṣduqu*) that A is half of C since half of half is not a half but a quarter." (Al-Jurjānī 2004, p. 153).

[25]*Al-Qiyās* II.4 in Hodges and Moktefi (2013, p. 36).

> Regarding the one in terms of equality, this is by virtue of some comparative similarity (*munāsaba*), in that the state of the ship, for example, with respect to the captain and the state of the city with respect to the king are one. (p. 78, § 17)

This example confirms our analysis that numbers and more generally classes of objects, which can be of completely different essences like ship and kingdom, can nevertheless be formed by the apprehension of a nonconcrete invariant namely the governance relation functioning as the similarity criterion, and as a result the ordered pair (captain, ship) and (king, city) forms an equivalence class (see Rahman and Salloum 2013, pp. 58–59). As mentioned in the first quote of this section equality is seen as numerical equality. Hence, it seems then that equality as a specific relation is defined in terms of the more general similar relation. The extensional contemporary definition is to identify equality as the smallest equivalence class. However, extensionality does not seem to be the conceptual frame of Ibn Sīnā. Perhaps the best way to think about it as establishing some kind of bijective function between two equivalence aggregates. At any rate, it very much looks as if Ibn Sīnā is attempting to elucidate equations such as 1/2 = 2/4 that are equal though strictly speaking they are very distinct expressions.

Chapter 6
Concluding Remarks

If there is one word that can sum up Ibn Sīnā's *al-Ilāhiyāt*, it is without doubt *intentionality*. It is present everywhere and it is intentionality that enables Ibn Sīnā to present science as a product of the creativity of the mind. He was so much impressed by the powerful creativity of the human mind that he wrote his famous poem *The Soul* in its praise, the belles-lettres of Arabic literature, to ensure the immortality of human mind's creative thought. He developed powerful arguments and subtle analyses to show us to what extent our mental life depends on intentionality: intentionality as mental existence, as reference to extra-mental entities as well as to previous experiences, as combinative unity by linking what otherwise seems separate (and, of course, to separate what otherwise seems indivisible), as the construction of new object by the discovery of a new meaning, as a predictive act by going beyond the actual intentional object. It has also an eminently epistemic dimension, which is of paramount importance to our author, for it is, with the support of memory, the basis of the mind's awareness of its own acts. The great originality of Ibn Sīnā's epistemology that could have transformed the philosophy of mathematics has been so far overlooked. There is still widespread prejudice against his works which ironically bear all the hallmarks of an author who was battling alone among those whom he described in his *Manṭiq al-Mashriqiyyīn* as "ignorants (الجهال)… devoid of understanding like propped up pieces of timber (عاري الفهم كأنهم خشب مسندة)" (Ibn Sīnā 1910, p. 3). He strove to go beyond Aristotle's logic and philosophy which were stagnating for centuries; sadly he ended up being reduced to an exegete of Aristotle. In fact, Ibn Sīnā's innovations emerged, as a result of the failure of a cooperative strategy by which he was hoping to smoothly bring change within the Aristotelian frame. The significance of *Manṭiq al-Mashriqiyyīn* is that it answers some puzzling attitudes of its author. His early Aristotelianism was a ploy, strategically used as an instrument to bring change to a tradition which was systematically opposed to innovation (التعمق في النظر بدعة) and creativity (و مخالفة المشهور ضلالة). These are revelations from someone who lived for a long time among the Aristotelians and are valuable insights for they give us an idea what it meant to work inside what seemed to be a very closed Greek tradition, as Aristotelianism was. His consistent bias in favour of the peripatetics (للمشائين تعصبنا) and his systematic defence of Aristotle and Aristotelian views was motivated by wanting to support their position, rightly or wrongly, against attacks from all the

H. Tahiri, *Mathematics and the Mind*,
SpringerBriefs in Philosophy, DOI 10.1007/978-3-319-25238-4_6

other traditions (mathematicians, scientists, jurists, *mutakallimūn*, linguists). One such important bias towards Aristotle was astutely identified by Rashed, it regards the classification of sciences in *al-Shifā'* (Rashed 2008, p. 164).

Ibn Sīnā's break with the Aristotelian tradition was provoked by the persistent resistance of this tradition to change. This was due to its inability to evolve with the new socio-epistemic context created by the great surge in the 8th century of new scientific disciplines which fostered "the major part of innovative research" (Rashed in al-Khwārizmī 2007, p. 16); for they gave rise to a "new scientific rationality which was essential to the elaboration of new scientific disciplines like algebra" where "*Épistèmè* and *Technè* do not exclude each other, and an apodictic knowledge can also have an intended target outside of it" (ibid., p. 18). His education was at odds with the formation of the Greek philosophical tradition as he had advanced mathematical knowledge. This enabled him to grasp the major impact on the philosophy of the ancients of the emergence of two powerful mathematical disciplines, algebra and arithmetic. His theory of knowledge, which is directly linked to the concept of number, is the result of extending the application of mathematics to metaphysics by linking what the Aristotelians have always inherited as separated; that is why he saw the need for logic and philosophy to change. Ibn Sīnā's experience is evidence that the Aristotelian tradition, which was adamant that its teacher had completed and perfected philosophy and virtually all the sciences, was a major obstacle to the development of science, mainly since the translation of Greek scientific and philosophical works. By losing its biggest supporter and protector, which Ibn Sīnā was to begin with, the powerful Aristotelian camp was badly weakened; its decline began and it found itself more vulnerable than ever to even sharper attacks to come. This brings us to another great misinterpretation of Arabic philosophy according to which al-Ghazālī's *Tahāfut al-falāsifa* has put an end to philosophy and its practice by demonstrating the inconsistency of Ibn Sīnā's system. The assumption is that following *Tahāfut* Ibn Sīnā's whole system has fallen apart at the seams. But al-Ghazālī only speaks of some local inconsistencies that he enumerated in his *Tahāfut*. These can be dissolved or at least isolated and as a consequence the system remains sound. In this sense, al-Ghazālī not only reinstated and validated substantial parts of Ibn Sīnā's system, he even contributed to its refinement.

If Ibn Sīnā's basic theory of knowledge has been broadly accepted by his posterity in the east as evidenced by al-Jurjānī's *Book of Definitions* which includes key concepts of his epistemology as he defined them, the reception of his works were mixed in the western Arab region. Though Ibn Bājja and Ibn Tofail were clearly influenced by some of Ibn Sīnā's epistemic views, none of his major works such as *al-Shifā'*, *al-Najāt* or *al-Ishārāt* have been comprehensively studied or specifically commented upon by western Arabic philosophers. As for Abūl-Walīd ibn Rushd, he can be best characterised as "anti-Ghazālī." Instead of considering al-Ghazālī's challenging arguments as some kind of awakening from his dogmatic slumber, he accuses his *Tahāfut* of disrupting and harming philosophy (Ibn Rushd 1998, p. 152) and blames as a consequence al-Fārābī's and Ibn Sīnā's innovations for making it vulnerable to al-Ghazālī's attacks by departing from Aristotle's teachings. As a result, he comes to the conclusion that philosophy needs little change. It appears that the philosopher of Cordoba has interpreted al-Ghazālī's *Tahāfut* as evidence that, with Aristotle, philosophy has reached its fixed point or, to use Ibn al-Haytham's terms, become an invariant belief; no significant change can make it in any way better. To counter the huge influence of al-Ghazālī's thought as he himself acknowledges by describing his advent as the "torrent that spills over into the villages" (ibid., p. 150), he committed himself once and for all to Aristotle's philosophical system from which he could never free himself. His hostility to the author of *Tahāfut al-falāsifa* is clear and unwavering: "that al-Ghazālī should touch on such questions in this way is not worthy of such a man: either he knew these matters in their true nature, and sets them out wrongly, which is wicked; or he did not understand their real nature and touched on problems he had not grasped, which is the act of an ignoramus." (Ibn Rushd 1930, p. 108). A serious misunderstanding that led the traditionalist philosopher to oppose and reject many of the innovations made by his predecessors since al-Kindī. The last Andalusian philosopher did not perceive that his traditional conception of philosophy was incompatible with the major development of many scientific disciplines that were further boosted by their interactions. The Commentator would be disappointed if he thought his various and comprehensive commentaries on Aristotle's works have put philosophy on the right track for he was followed neither in the east nor in the west. His negative attitude towards Arabic and Islamic philosophers and almost total neglect of their works has two major implications: unlike the eastern Arabic region, philosophy, that he portrayed as closed and complete system, has stagnated once again this time in the western part of the Arabic-Islamic world. The Scholastics, who did not and could not know the story of the development of Arabic philosophy that was unfolding, were particularly deprived of getting access to Ibn Sīnā's vast works in which he fully developed his philosophical system. Though the bulk of Arabic philosophical writings that reached the Scholastics was Ibn Rushd's monumental commentary works, the Scholastics did not and could not share the Commentator's conception of philosophy as an eternal exegesis even if they had wanted to because of a simple reason: the growing transformation of science led by the unstoppable expansion of mathematics effectively completed the de-hellinisation of the Greek conception of science and philosophy.

As for modern scholars, it is unfortunate that most of them, who almost systematically link Ibn Sīnā's views backwards to the pre-al-Khwārizmī era, mainly to Aristotle and the Aristotelian tradition, as if he had never publicly declared the contrary, continue to neglect his mathematical writings; while, in our view, his epistemology of mathematics as expounded in *al-Ilāhiyāt*, which opens a new chapter in the history of philosophy, irresistibly drives us forward to 19th and 20th centuries' research on the foundations of mathematics. For it remarkably turns out that Ibn Sīnā's work on the philosophy of mathematics strongly reminds us of similar works undertaken by Husserl and even by contemporary constructivists. And yet there is no epistemological study that links Husserl's intentionality, a powerful notion that enables him alone to found a new philosophical tradition, with its origin in the Arabic tradition. It is the similarities of contemporary and Arabic approaches that make the comparison not only historically instructive but also theoretically useful to better appreciate the contribution of Ibn Sīnā to issues relating to the foundations of mathematics. But that is another story; as for the present, it is just another pointer to and reminder of the gaps in the history of the development of science.

References

Primary Sources

Works of Ibn Sīnā

Avicenna (2005) The metaphysics of the healing (الشفاء: الإلهيات). A parallel English-Arabic text translated, introduced and annotated by Michael E. Marmura. Islamic Translation Series. Birgham Young University Press, Prove, Utah

Avicenna (2009) The physics of the healing (الشفاء: السماع الطبيعي). A parallel English-Arabic text translated, introduced and annotated by Jon McGinnis. Islamic translation series, Birgham Young University Press, Prove, Utah

Avicenne (1951) LIVRE DES DIRECTIVES ET REMARQUES. KITĀB AL-'ISHĀRĀT WA L-TANBĪHĀT, Traduction, introduction et notes par A.-M. Goichon, Paris-Beyrouth, Editions UNESCO, Vrin

Avicenne (1978) La Métaphysique du Shifā', Livres I à V, Introduction, traduction et notes par Georges C. Anawati, J. Vrin, Paris

Avicenne (1986) Le Livre de Science. Traduit du persan par Mohammad Achena et Henri Massé, Deuxième édition revue et corrigée, Les Belles Lettres–UNESCO

Gohlman WE (1974) The life of Ibn Sina. A critical edition and annotated translation. In: Studies in Islamic Philosophy and Science (Trans: Gohlman WE). State University of New York Press, Albany

Ibn Sīnā (1910) Manṭiq al-mashriqiyyīn, ed. al-Maktaba al-Salafiyya, Maṭbaʿat al-Murīd, Cairo

Ibn Sīnā (1956) Al-Shifā', al-Manṭiq, vol V: *al-Burhān*, ed. Afifi, directed by Ibrahim Madhkour, al-Maṭbaʿa al-Amīrīa, Cairo

Ibn Sīnā (1968) Kitāb al-ishārāt wa al-tanbīhāt with a commentary by Nasīr al-Dīn al-Ṭūsī, Third Part, ed. S. Dunyā, Dār al-Ma'ārif, Cairo

Ibn Sīnā (1983) Kitāb al-ishārāt wa al-tanbīhāt with a commentary by Nasīr al-Dīn al-Ṭūsī, First Part One, ed. S. Dunyā, Dār al-Ma'ārif, Cairo

Ibn Sīnā (1992) Kitāb al-ishārāt wa al-tanbīhāt with a commentary by Nasīr al-Dīn al-Ṭūsī, Second Part, ed. S. Dunyā, Dār al-Ma'ārif, Cairo

Ibn Sīnā (1989) Tisʿ rasā'il fī al-ḥikma wa al-ṭabīʿiyyāt. Dār al-ʿArab lil-Bustānī, Cairo

H. Tahiri, *Mathematics and the Mind*,
SpringerBriefs in Philosophy, DOI 10.1007/978-3-319-25238-4

Other Works

Al-Bājī, Abū al-Walīd (1978), *Kitāb al-minhāj fī tartīb al-ḥijāj*, ed. A. Turki. Paris, Maisonneuve et Larose

Al-Fārābī (1890) Fī Aghrāḍ kitāb mā ba'da aṭ-ṭabī'a. In: Dieterici F (ed) Al-Fārābī's Philosophische Abhandlungen. Brill, Leiden, pp 34–38

Al-Fārābī (1970) Kitāb al-ḥurūf, ed. Muhsin Mahdi, Beirut

Al-Fārābī (1989) Kitāb al-wāḥid wa al-waḥda. Arabic text edited with introduction and notes by M. Mahdi. Casablanca, Ṭūbqāl li al-Nashr

Al-Fārābī (1996) Iḥsā' al-'ulūm. Dār wa Maktabat al-Hilāl, Beirut

Al-Ghazālī (1961) Maqāṣid al-falāsifa, ed. S. Dunya, Dhakhā'iru al-'Arab 29, Dār al-Ma'ārif bi Miṣr

Al-Ghazālī (2002) Tahāfut al-falāsifa (The incoherence of the philosophers). A parallel English-Arabic text translated, introduced and annotated by Michael E. Marmura, Islamic translation series. Birgham Young University Press, Prove, Utah

Al-Jurjānī (2004) Mu'jam al-ta'rīfāt, ed. Muhammad Siddiq al-Minshawi. Dār al-Faḍīla, Cairo

Al-Juwainī, Imām al-Ḥaramain (1979) Al-Kāfiya fī al-jadal, ed. F. Husain Mahmud, Cairo, Maṭba'at 'Isā al-Ḥalabī

Al-Khwārizmī (2007) Le commencement de l'algèbre, texte établi, traduit et commenté par R. Rashed, Albert Blanchard, Paris; for the English version Al-Khwārizmī: The Beginnings of Algebra, Edited with translation and commentary by Roshdi Rashed, Bilingual edition, Saqi Books, 2009

Annas J (1976) Aristotle's metaphysics. Books M and N. Translated with introduction and notes by Julia Annas, Oxford University Press, Oxford

Grant E (1974) A sourcebook in medieval science. Source books in the history of the sciences (Book 13), Harvard University Press

Hodges W, Moktefi A (2013) Ibn Sīnā on logical analysis, (Draft). Available at Hodges's website: http://wilfridhodges.co.uk/arabic29.pdf

Ḥunain ibn Isḥaq (1985) Ādāb al-falāsifa, edited and introduced by A. Badawi, Ma'had al-Makhṭūṭāt al-'Arabiyya, al-Munaẓẓama al-'Arabiyya lit-Tarbiya wa al-Thaqāfa wa al-'Ulūm, al-Ṣafāt, al-Kuwait

Ibn Fāris (1366/1984) Mu'jam maqāyīs al-lugha, ed. A. Harun, vol 6, Cairo

Ibn al-'Imād: Shadharāt al-dhahab fī akhbār man dhahab, vol 3, Beirut (no date)

Ibn al-Haytham (1971) Al-Shukūk 'ala Baṭlamyūs, ed. A.I. Sabra and N. Shehaby. Dar al-Kutub, Cairo

Ibn Khallikān (1969) Wafayāt al-a'yān, vol 2, ed. Iḥsān 'Abbās, Beirut

Ibn Rushd (1930) Tahāfut al-tahāfut, ed. M. Bouyges, Beirut, Imprimerie Catholique

Ibn Rushd (1954) Averroes' Tahāfut al-Tahāfut (The Incoherence of the Incoherence), Translated by S. van den Bergh in 2 Volumes: vol I, pp 374, vol II, pp 219, Gibb Memorial Series, N.S. XIX, Oxford University Press

Ibn Rushd (1958) Talkhīṣ mā ba'da al-ṭabī'a, ed. 'Uthman Amin, Cairo

Ibn Rushd (1998) Al-Kashf 'an manāhij al-'adilla fī 'aqā'id al-milla. Markaz Dirāsat al-Waḥda al-'Arabiyya, Silsilat al-Turāth al-Falsafi al-'Arabī, Mu'allafāt Ibn Rushd (2), Beirut

Ross WD (1924) Aristotle's metaphysics, a revised text with introduction and a commentary. Clarendon press, Oxford

Ross WD (1949) Aristotle's prior and posterior analytics, a revised text with introduction and commentary. Clarendon Press, Oxford

Secondary Literature

Badawi A (1978) Aristū 'inda al-'arab. al-Kuwait, Wakālat al-Maṭbū'at

Brentano F (1995) Psychology from an empirical standpoint. Originally published in 1874. English edition: McAlister L (1973), Routledge and Kegan Paul, London. Reprinted with an introduction by Peter Simons, Routledge, London

Burnet J (1908) Early Greek philosophy, 2nd edn. Adam and Charles Black, London

Crubellier M (1997) Platon, les nombres et Aristote. In: La mémoire des nombres. Actes du 10ème colloque inter-IREM d'épistémologie & d'histoire des mathématiques - université de Caen - Cherbourg (1994), IREM de Basse-Normandie, Caen, pp 81–100

Duhem P (1913–1965) Le Système du monde: Histoire des doctrines cosmologiques de Platon à Copernic, 10 volumes, Paris, Librairie scientifique A. Hermann et fils

Farmer HG (1932) The influence of Al-Fārābī's "Iḥsā' al-'ulūm" ("De scientiis") on the writers on music in Western Europe. J Roy Asiatic Soci G B Irel 3:561–592

Frege G (1884) The foundations of arithmetic, a logico-mathematical enquiry into the concept of number (Trans: Austin JL), 2nd revised edn, Harper Torchbooks. The Science Library Harper & Brothers, New York

Gauss CF (1900) Werke, vol VIII, Arithmetik und Algebra: Nachträge zu Band 1-3, ed. Königliche Gesellschaft der Wissenschaften zu Göttingen, Leipzig: Teubner; Repr. Hildesheim, New York: Olms, 1981

Kamlah W, Lorenzen P (1984) Logical propaedeutic, Translation of Logische Propädeutik (1972) by H. Robinson, University Press of America

King DA (1993) Astronomy in the service of Islam. Variorum collected studies series: CS416, Ashgate Variorum

Lasserre F (1990) La naissance des mathématiques à l'époque de Platon. Editions Universitaires Fribourg Suisse, Editions du Cerf Paris

Martin-Löf P (1984) Intuitionistic type theory—notes by Giovanni Sambin of a series of lectures given in Padua, June 1980. Bibliopolis, Naples

Miller LB (1985) Islamic disputation theory: a study of the development of dialectic in Islam from the tenth through fourteenth centuries. Princeton University (Unpublished dissertation)

Poincaré H (1902) La science et l'hypothèse. Flammarion, Paris

Poincaré H (1905) Les mathématiques et la logique. In: Revue de Métaphysique et de Morale, vol 13, pp 815–835; repris dans Science et Méthode (livre II, chapitre III). Flammarion, Paris, 1908

Prakken H, Vreeswijk G (2002) Logics for defeasible argumentation. In: Gabbay DM, Guenthner F (eds) Handbook of philosophical logic, 2nd edn, vol 4. Springer, Berlin, pp 219–318

Rahman S, Salloum Z (2013) Quantité, unité et identité dans la philosophie d'Ibn Sīnā. In: Tahiri H (ed) La périodisation en histoire des sciences et de la philosophie. La fin d'un mythe. Cahiers de Logique et d'Epistémologie, vol 13. College Publications, London, pp 51–61

Rahman S, Street T, Tahiri H (eds) (2008) The unity of science in the Arabic tradition: science, logic, epistemology and their interactions. Springer, Berlin

Rahman S, Granström JG, Salloum Z (2014) Ibn Sīnā's approach to equality and unity. Arabic Sci Philos 24:297–307

Rahsed R (1978) "La notion de science occidentale" In: E.G. Forbes (ed) Human implications of scientific advance, proceedings of the XVth international congress of the history of science, Edinburgh, Edinburgh University Press, pp. 45–54. [English trans: Science as a Western Phenomenon. In: Fundamenta Scientiae, 1 (1980), pp. 7–21. Arabic trans: In: Al-Mustaqbal al Arabī, 47 (1983), pp. 4–19]

Rashed R (2002) Les mathématiques infinitésimales du IXe au XIe. Volume IV Ibn al-Haytham: Méthodes géométriques, Transformations ponctuelles et Philosophie des mathématiques. al-Furqān Islamic Heritage Foundation, London

Rashed R (2008) Philosophy of mathematics. In: Rahman S, Street T, Tahiri H (eds) The unity of science in the arabic tradition: science: logic, epistemology and their interactions, vol 11. Springer, Berlin, pp 153–182

Rosenthal F (1970) Knowledge triumphant. The concept of knowledge in Medieval Islam, EJ Brill, Leiden

Ross D (1995) Aristotle, 6th edn. Routledge, London

Tahiri H (2008) The birth of scientific controversies: the dynamic of the Arabic tradition and its impact on the development of science: Ibn al-Haytham's challenge of Ptolemy's Almagest. In: Rahman S, Street T, Tahiri H (eds) The unity of science in the Arabic tradition: science, logic, epistemology and their interactions. Springer, Berlin, pp 183–225

Van der Waerden BL (1961) Science awakening (English trans: Dresden A with additions of the author). Oxford University Press, New York

MIX
Papier aus verantwortungsvollen Quellen
Paper from responsible sources
FSC® C105338

If you have any concerns about our products,
you can contact us on
ProductSafety@springernature.com

In case Publisher is established outside the EU,
the EU authorized representative is:
Springer Nature Customer Service Center GmbH
Europaplatz 3, 69115 Heidelberg, Germany

Printed by Libri Plureos GmbH
in Hamburg, Germany